Topology: A Very Short Introduction

VERY SHORT INTRODUCTIONS are for anyone wanting a stimulating and accessible way into a new subject. They are written by experts, and have been translated into more than 45 different languages.

The series began in 1995, and now covers a wide variety of topics in every discipline. The VSI library currently contains over 600 volumes—a Very Short Introduction to everything from Psychology and Philosophy of Science to American History and Relativity—and continues to grow in every subject area.

Very Short Introductions available now:

ABOLITIONISM Richard S. Newman
THE ABRAHAMIC RELIGIONS Charles L. Cohen
ACCOUNTING Christopher Nobes
ADAM SMITH Christopher J. Berry
ADOLESCENCE Peter K. Smith
ADVERTISING Winston Fletcher
AESTHETICS Bence Nanay
AFRICAN AMERICAN RELIGION Eddie S. Glaude Jr
AFRICAN HISTORY John Parker and Richard Rathbone
AFRICAN POLITICS Ian Taylor
AFRICAN RELIGIONS Jacob K. Olupona
AGEING Nancy A. Pachana
AGNOSTICISM Robin Le Poidevin
AGRICULTURE Paul Brassley and Richard Soffe
ALEXANDER THE GREAT Hugh Bowden
ALGEBRA Peter M. Higgins
AMERICAN CULTURAL HISTORY Eric Avila
AMERICAN FOREIGN RELATIONS Andrew Preston
AMERICAN HISTORY Paul S. Boyer
AMERICAN IMMIGRATION David A. Gerber
AMERICAN LEGAL HISTORY G. Edward White
AMERICAN NAVAL HISTORY Craig L. Symonds
AMERICAN POLITICAL HISTORY Donald Critchlow
AMERICAN POLITICAL PARTIES AND ELECTIONS L. Sandy Maisel
AMERICAN POLITICS Richard M. Valelly
THE AMERICAN PRESIDENCY Charles O. Jones
THE AMERICAN REVOLUTION Robert J. Allison
AMERICAN SLAVERY Heather Andrea Williams
THE AMERICAN WEST Stephen Aron
AMERICAN WOMEN'S HISTORY Susan Ware
ANAESTHESIA Aidan O'Donnell
ANALYTIC PHILOSOPHY Michael Beaney
ANARCHISM Colin Ward
ANCIENT ASSYRIA Karen Radner
ANCIENT EGYPT Ian Shaw
ANCIENT EGYPTIAN ART AND ARCHITECTURE Christina Riggs
ANCIENT GREECE Paul Cartledge
THE ANCIENT NEAR EAST Amanda H. Podany
ANCIENT PHILOSOPHY Julia Annas
ANCIENT WARFARE Harry Sidebottom
ANGELS David Albert Jones
ANGLICANISM Mark Chapman
THE ANGLO-SAXON AGE John Blair
ANIMAL BEHAVIOUR Tristram D. Wyatt
THE ANIMAL KINGDOM Peter Holland
ANIMAL RIGHTS David DeGrazia

THE ANTARCTIC Klaus Dodds
ANTHROPOCENE Erle C. Ellis
ANTISEMITISM Steven Beller
ANXIETY Daniel Freeman and Jason Freeman
THE APOCRYPHAL GOSPELS Paul Foster
APPLIED MATHEMATICS Alain Goriely
ARCHAEOLOGY Paul Bahn
ARCHITECTURE Andrew Ballantyne
ARISTOCRACY William Doyle
ARISTOTLE Jonathan Barnes
ART HISTORY Dana Arnold
ART THEORY Cynthia Freeland
ARTIFICIAL INTELLIGENCE Margaret A. Boden
ASIAN AMERICAN HISTORY Madeline Y. Hsu
ASTROBIOLOGY David C. Catling
ASTROPHYSICS James Binney
ATHEISM Julian Baggini
THE ATMOSPHERE Paul I. Palmer
AUGUSTINE Henry Chadwick
AUSTRALIA Kenneth Morgan
AUTISM Uta Frith
AUTOBIOGRAPHY Laura Marcus
THE AVANT GARDE David Cottington
THE AZTECS Davíd Carrasco
BABYLONIA Trevor Bryce
BACTERIA Sebastian G. B. Amyes
BANKING John Goddard and John O. S. Wilson
BARTHES Jonathan Culler
THE BEATS David Sterritt
BEAUTY Roger Scruton
BEHAVIOURAL ECONOMICS Michelle Baddeley
BESTSELLERS John Sutherland
THE BIBLE John Riches
BIBLICAL ARCHAEOLOGY Eric H. Cline
BIG DATA Dawn E. Holmes
BIOGRAPHY Hermione Lee
BIOMETRICS Michael Fairhurst
BLACK HOLES Katherine Blundell
BLOOD Chris Cooper
THE BLUES Elijah Wald
THE BODY Chris Shilling
THE BOOK OF COMMON PRAYER Brian Cummings
THE BOOK OF MORMON Terryl Givens
BORDERS Alexander C. Diener and Joshua Hagen
THE BRAIN Michael O'Shea
BRANDING Robert Jones
THE BRICS Andrew F. Cooper
THE BRITISH CONSTITUTION Martin Loughlin
THE BRITISH EMPIRE Ashley Jackson
BRITISH POLITICS Anthony Wright
BUDDHA Michael Carrithers
BUDDHISM Damien Keown
BUDDHIST ETHICS Damien Keown
BYZANTIUM Peter Sarris
C. S. LEWIS James Como
CALVINISM Jon Balserak
CANCER Nicholas James
CAPITALISM James Fulcher
CATHOLICISM Gerald O'Collins
CAUSATION Stephen Mumford and Rani Lill Anjum
THE CELL Terence Allen and Graham Cowling
THE CELTS Barry Cunliffe
CHAOS Leonard Smith
CHARLES DICKENS Jenny Hartley
CHEMISTRY Peter Atkins
CHILD PSYCHOLOGY Usha Goswami
CHILDREN'S LITERATURE Kimberley Reynolds
CHINESE LITERATURE Sabina Knight
CHOICE THEORY Michael Allingham
CHRISTIAN ART Beth Williamson
CHRISTIAN ETHICS D. Stephen Long
CHRISTIANITY Linda Woodhead
CIRCADIAN RHYTHMS Russell Foster and Leon Kreitzman
CITIZENSHIP Richard Bellamy
CIVIL ENGINEERING David Muir Wood
CLASSICAL LITERATURE William Allan
CLASSICAL MYTHOLOGY Helen Morales
CLASSICS Mary Beard and John Henderson
CLAUSEWITZ Michael Howard
CLIMATE Mark Maslin
CLIMATE CHANGE Mark Maslin
CLINICAL PSYCHOLOGY Susan Llewelyn and Katie Aafjes-van Doorn

COGNITIVE NEUROSCIENCE Richard Passingham
THE COLD WAR Robert McMahon
COLONIAL AMERICA Alan Taylor
COLONIAL LATIN AMERICAN LITERATURE Rolena Adorno
COMBINATORICS Robin Wilson
COMEDY Matthew Bevis
COMMUNISM Leslie Holmes
COMPARATIVE LITERATURE Ben Hutchinson
COMPLEXITY John H. Holland
THE COMPUTER Darrel Ince
COMPUTER SCIENCE Subrata Dasgupta
CONCENTRATION CAMPS Dan Stone
CONFUCIANISM Daniel K. Gardner
THE CONQUISTADORS Matthew Restall and Felipe Fernández-Armesto
CONSCIENCE Paul Strohm
CONSCIOUSNESS Susan Blackmore
CONTEMPORARY ART Julian Stallabrass
CONTEMPORARY FICTION Robert Eaglestone
CONTINENTAL PHILOSOPHY Simon Critchley
COPERNICUS Owen Gingerich
CORAL REEFS Charles Sheppard
CORPORATE SOCIAL RESPONSIBILITY Jeremy Moon
CORRUPTION Leslie Holmes
COSMOLOGY Peter Coles
COUNTRY MUSIC Richard Carlin
CRIME FICTION Richard Bradford
CRIMINAL JUSTICE Julian V. Roberts
CRIMINOLOGY Tim Newburn
CRITICAL THEORY Stephen Eric Bronner
THE CRUSADES Christopher Tyerman
CRYPTOGRAPHY Fred Piper and Sean Murphy
CRYSTALLOGRAPHY A. M. Glazer
THE CULTURAL REVOLUTION Richard Curt Kraus
DADA AND SURREALISM David Hopkins
DANTE Peter Hainsworth and David Robey
DARWIN Jonathan Howard
THE DEAD SEA SCROLLS Timothy H. Lim
DECADENCE David Weir
DECOLONIZATION Dane Kennedy
DEMOCRACY Bernard Crick
DEMOGRAPHY Sarah Harper
DEPRESSION Jan Scott and Mary Jane Tacchi
DERRIDA Simon Glendinning
DESCARTES Tom Sorell
DESERTS Nick Middleton
DESIGN John Heskett
DEVELOPMENT Ian Goldin
DEVELOPMENTAL BIOLOGY Lewis Wolpert
THE DEVIL Darren Oldridge
DIASPORA Kevin Kenny
DICTIONARIES Lynda Mugglestone
DINOSAURS David Norman
DIPLOMACY Joseph M. Siracusa
DOCUMENTARY FILM Patricia Aufderheide
DREAMING J. Allan Hobson
DRUGS Les Iversen
DRUIDS Barry Cunliffe
DYNASTY Jeroen Duindam
DYSLEXIA Margaret J. Snowling
EARLY MUSIC Thomas Forrest Kelly
THE EARTH Martin Redfern
EARTH SYSTEM SCIENCE Tim Lenton
ECONOMICS Partha Dasgupta
EDUCATION Gary Thomas
EGYPTIAN MYTH Geraldine Pinch
EIGHTEENTH-CENTURY BRITAIN Paul Langford
THE ELEMENTS Philip Ball
EMOTION Dylan Evans
EMPIRE Stephen Howe
ENERGY SYSTEMS Nick Jenkins
ENGELS Terrell Carver
ENGINEERING David Blockley
THE ENGLISH LANGUAGE Simon Horobin
ENGLISH LITERATURE Jonathan Bate
THE ENLIGHTENMENT John Robertson
ENTREPRENEURSHIP Paul Westhead and Mike Wright
ENVIRONMENTAL ECONOMICS Stephen Smith

ENVIRONMENTAL ETHICS
Robin Attfield
ENVIRONMENTAL LAW
Elizabeth Fisher
ENVIRONMENTAL POLITICS
Andrew Dobson
EPICUREANISM Catherine Wilson
EPIDEMIOLOGY Rodolfo Saracci
ETHICS Simon Blackburn
ETHNOMUSICOLOGY Timothy Rice
THE ETRUSCANS Christopher Smith
EUGENICS Philippa Levine
THE EUROPEAN UNION
Simon Usherwood and John Pinder
EUROPEAN UNION LAW
Anthony Arnull
EVOLUTION Brian and
Deborah Charlesworth
EXISTENTIALISM Thomas Flynn
EXPLORATION Stewart A. Weaver
EXTINCTION Paul B. Wignall
THE EYE Michael Land
FAIRY TALE Marina Warner
FAMILY LAW Jonathan Herring
FASCISM Kevin Passmore
FASHION Rebecca Arnold
FEDERALISM Mark J. Rozell and
Clyde Wilcox
FEMINISM Margaret Walters
FILM Michael Wood
FILM MUSIC Kathryn Kalinak
FILM NOIR James Naremore
THE FIRST WORLD WAR
Michael Howard
FOLK MUSIC Mark Slobin
FOOD John Krebs
FORENSIC PSYCHOLOGY
David Canter
FORENSIC SCIENCE Jim Fraser
FORESTS Jaboury Ghazoul
FOSSILS Keith Thomson
FOUCAULT Gary Gutting
THE FOUNDING FATHERS
R. B. Bernstein
FRACTALS Kenneth Falconer
FREE SPEECH Nigel Warburton
FREE WILL Thomas Pink
FREEMASONRY Andreas Önnerfors
FRENCH LITERATURE John D. Lyons
THE FRENCH REVOLUTION
William Doyle
FREUD Anthony Storr
FUNDAMENTALISM Malise Ruthven
FUNGI Nicholas P. Money
THE FUTURE Jennifer M. Gidley
GALAXIES John Gribbin
GALILEO Stillman Drake
GAME THEORY Ken Binmore
GANDHI Bhikhu Parekh
GARDEN HISTORY Gordon Campbell
GENES Jonathan Slack
GENIUS Andrew Robinson
GENOMICS John Archibald
GEOFFREY CHAUCER David Wallace
GEOGRAPHY JOHN Matthews and
David Herbert
GEOLOGY Jan Zalasiewicz
GEOPHYSICS William Lowrie
GEOPOLITICS Klaus Dodds
GERMAN LITERATURE Nicholas Boyle
GERMAN PHILOSOPHY
Andrew Bowie
GLACIATION David J. A. Evans
GLOBAL CATASTROPHES Bill McGuire
GLOBAL ECONOMIC HISTORY
Robert C. Allen
GLOBALIZATION Manfred Steger
GOD John Bowker
GOETHE Ritchie Robertson
THE GOTHIC Nick Groom
GOVERNANCE Mark Bevir
GRAVITY Timothy Clifton
THE GREAT DEPRESSION AND
THE NEW DEAL Eric Rauchway
HABERMAS James Gordon Finlayson
THE HABSBURG EMPIRE
Martyn Rady
HAPPINESS Daniel M. Haybron
THE HARLEM RENAISSANCE
Cheryl A. Wall
THE HEBREW BIBLE AS LITERATURE
Tod Linafelt
HEGEL Peter Singer
HEIDEGGER Michael Inwood
THE HELLENISTIC AGE
Peter Thonemann
HEREDITY John Waller
HERMENEUTICS Jens Zimmermann
HERODOTUS Jennifer T. Roberts
HIEROGLYPHS Penelope Wilson
HINDUISM Kim Knott
HISTORY John H. Arnold

THE HISTORY OF ASTRONOMY
Michael Hoskin
THE HISTORY OF CHEMISTRY
William H. Brock
THE HISTORY OF CHILDHOOD
James Marten
THE HISTORY OF CINEMA
Geoffrey Nowell-Smith
THE HISTORY OF LIFE
Michael Benton
THE HISTORY OF MATHEMATICS
Jacqueline Stedall
THE HISTORY OF MEDICINE
William Bynum
THE HISTORY OF PHYSICS
J. L. Heilbron
THE HISTORY OF TIME
Leofranc Holford-Strevens
HIV AND AIDS Alan Whiteside
HOBBES Richard Tuck
HOLLYWOOD Peter Decherney
THE HOLY ROMAN EMPIRE
Joachim Whaley
HOME Michael Allen Fox
HOMER Barbara Graziosi
HORMONES Martin Luck
HUMAN ANATOMY
Leslie Klenerman
HUMAN EVOLUTION Bernard Wood
HUMAN RIGHTS Andrew Clapham
HUMANISM Stephen Law
HUME A. J. Ayer
HUMOUR Noël Carroll
THE ICE AGE Jamie Woodward
IDENTITY Florian Coulmas
IDEOLOGY Michael Freeden
THE IMMUNE SYSTEM
Paul Klenerman
INDIAN CINEMA
Ashish Rajadhyaksha
INDIAN PHILOSOPHY Sue Hamilton
THE INDUSTRIAL REVOLUTION
Robert C. Allen
INFECTIOUS DISEASE Marta L. Wayne
and Benjamin M. Bolker
INFINITY Ian Stewart
INFORMATION Luciano Floridi
INNOVATION Mark Dodgson
and David Gann
INTELLECTUAL PROPERTY
Siva Vaidhyanathan
INTELLIGENCE Ian J. Deary
INTERNATIONAL LAW
Vaughan Lowe
INTERNATIONAL MIGRATION
Khalid Koser
INTERNATIONAL RELATIONS
Paul Wilkinson
INTERNATIONAL SECURITY
Christopher S. Browning
IRAN Ali M. Ansari
ISLAM Malise Ruthven
ISLAMIC HISTORY Adam Silverstein
ISOTOPES Rob Ellam
ITALIAN LITERATURE
Peter Hainsworth and David Robey
JESUS Richard Bauckham
JEWISH HISTORY David N. Myers
JOURNALISM Ian Hargreaves
JUDAISM Norman Solomon
JUNG Anthony Stevens
KABBALAH Joseph Dan
KAFKA Ritchie Robertson
KANT Roger Scruton
KEYNES Robert Skidelsky
KIERKEGAARD Patrick Gardiner
KNOWLEDGE Jennifer Nagel
THE KORAN Michael Cook
LAKES Warwick F. Vincent
LANDSCAPE ARCHITECTURE
Ian H. Thompson
LANDSCAPES AND
GEOMORPHOLOGY
Andrew Goudie and Heather Viles
LANGUAGES Stephen R. Anderson
LATE ANTIQUITY Gillian Clark
LAW Raymond Wacks
THE LAWS OF THERMODYNAMICS
Peter Atkins
LEADERSHIP Keith Grint
LEARNING Mark Haselgrove
LEIBNIZ Maria Rosa Antognazza
LEO TOLSTOY Liza Knapp
LIBERALISM Michael Freeden
LIGHT Ian Walmsley
LINCOLN Allen C. Guelzo
LINGUISTICS Peter Matthews
LITERARY THEORY Jonathan Culler
LOCKE John Dunn
LOGIC Graham Priest
LOVE Ronald de Sousa
MACHIAVELLI Quentin Skinner

MADNESS Andrew Scull
MAGIC Owen Davies
MAGNA CARTA Nicholas Vincent
MAGNETISM Stephen Blundell
MALTHUS Donald Winch
MAMMALS T. S. Kemp
MANAGEMENT John Hendry
MAO Delia Davin
MARINE BIOLOGY Philip V. Mladenov
THE MARQUIS DE SADE John Phillips
MARTIN LUTHER Scott H. Hendrix
MARTYRDOM Jolyon Mitchell
MARX Peter Singer
MATERIALS Christopher Hall
MATHEMATICAL FINANCE Mark H. A. Davis
MATHEMATICS Timothy Gowers
MATTER Geoff Cottrell
THE MEANING OF LIFE Terry Eagleton
MEASUREMENT David Hand
MEDICAL ETHICS Michael Dunn and Tony Hope
MEDICAL LAW Charles Foster
MEDIEVAL BRITAIN John Gillingham and Ralph A. Griffiths
MEDIEVAL LITERATURE Elaine Treharne
MEDIEVAL PHILOSOPHY John Marenbon
MEMORY Jonathan K. Foster
METAPHYSICS Stephen Mumford
METHODISM William J. Abraham
THE MEXICAN REVOLUTION Alan Knight
MICHAEL FARADAY Frank A. J. L. James
MICROBIOLOGY Nicholas P. Money
MICROECONOMICS Avinash Dixit
MICROSCOPY Terence Allen
THE MIDDLE AGES Miri Rubin
MILITARY JUSTICE Eugene R. Fidell
MILITARY STRATEGY Antulio J. Echevarria II
MINERALS David Vaughan
MIRACLES Yujin Nagasawa
MODERN ARCHITECTURE Adam Sharr
MODERN ART David Cottington
MODERN CHINA Rana Mitter
MODERN DRAMA Kirsten E. Shepherd-Barr
MODERN FRANCE Vanessa R. Schwartz
MODERN INDIA Craig Jeffrey
MODERN IRELAND Senia Pašeta
MODERN ITALY Anna Cento Bull
MODERN JAPAN Christopher Goto-Jones
MODERN LATIN AMERICAN LITERATURE Roberto González Echevarría
MODERN WAR Richard English
MODERNISM Christopher Butler
MOLECULAR BIOLOGY Aysha Divan and Janice A. Royds
MOLECULES Philip Ball
MONASTICISM Stephen J. Davis
THE MONGOLS Morris Rossabi
MOONS David A. Rothery
MORMONISM Richard Lyman Bushman
MOUNTAINS Martin F. Price
MUHAMMAD Jonathan A. C. Brown
MULTICULTURALISM Ali Rattansi
MULTILINGUALISM John C. Maher
MUSIC Nicholas Cook
MYTH Robert A. Segal
NAPOLEON David Bell
THE NAPOLEONIC WARS Mike Rapport
NATIONALISM Steven Grosby
NATIVE AMERICAN LITERATURE Sean Teuton
NAVIGATION Jim Bennett
NAZI GERMANY Jane Caplan
NELSON MANDELA Elleke Boehmer
NEOLIBERALISM Manfred Steger and Ravi Roy
NETWORKS Guido Caldarelli and Michele Catanzaro
THE NEW TESTAMENT Luke Timothy Johnson
THE NEW TESTAMENT AS LITERATURE Kyle Keefer
NEWTON Robert Iliffe
NIETZSCHE Michael Tanner
NINETEENTH-CENTURY BRITAIN Christopher Harvie and H. C. G. Matthew

THE NORMAN CONQUEST George Garnett
NORTH AMERICAN INDIANS Theda Perdue and Michael D. Green
NORTHERN IRELAND Marc Mulholland
NOTHING Frank Close
NUCLEAR PHYSICS Frank Close
NUCLEAR POWER Maxwell Irvine
NUCLEAR WEAPONS Joseph M. Siracusa
NUMBERS Peter M. Higgins
NUTRITION David A. Bender
OBJECTIVITY Stephen Gaukroger
OCEANS Dorrik Stow
THE OLD TESTAMENT Michael D. Coogan
THE ORCHESTRA D. Kern Holoman
ORGANIC CHEMISTRY Graham Patrick
ORGANIZATIONS Mary Jo Hatch
ORGANIZED CRIME Georgios A. Antonopoulos and Georgios Papanicolaou
ORTHODOX CHRISTIANITY A. Edward Siecienski
PAGANISM Owen Davies
PAIN Rob Boddice
THE PALESTINIAN-ISRAELI CONFLICT Martin Bunton
PANDEMICS Christian W. McMillen
PARTICLE PHYSICS Frank Close
PAUL E. P. Sanders
PEACE Oliver P. Richmond
PENTECOSTALISM William K. Kay
PERCEPTION Brian Rogers
THE PERIODIC TABLE Eric R. Scerri
PHILOSOPHY Edward Craig
PHILOSOPHY IN THE ISLAMIC WORLD Peter Adamson
PHILOSOPHY OF BIOLOGY Samir Okasha
PHILOSOPHY OF LAW Raymond Wacks
PHILOSOPHY OF SCIENCE Samir Okasha
PHILOSOPHY OF RELIGION Tim Bayne
PHOTOGRAPHY Steve Edwards
PHYSICAL CHEMISTRY Peter Atkins
PHYSICS Sidney Perkowitz
PILGRIMAGE Ian Reader
PLAGUE Paul Slack
PLANETS David A. Rothery
PLANTS Timothy Walker
PLATE TECTONICS Peter Molnar
PLATO Julia Annas
POETRY Bernard O'Donoghue
POLITICAL PHILOSOPHY David Miller
POLITICS Kenneth Minogue
POPULISM Cas Mudde and Cristóbal Rovira Kaltwasser
POSTCOLONIALISM Robert Young
POSTMODERNISM Christopher Butler
POSTSTRUCTURALISM Catherine Belsey
POVERTY Philip N. Jefferson
PREHISTORY Chris Gosden
PRESOCRATIC PHILOSOPHY Catherine Osborne
PRIVACY Raymond Wacks
PROBABILITY John Haigh
PROGRESSIVISM Walter Nugent
PROJECTS Andrew Davies
PROTESTANTISM Mark A. Noll
PSYCHIATRY Tom Burns
PSYCHOANALYSIS Daniel Pick
PSYCHOLOGY Gillian Butler and Freda McManus
PSYCHOLOGY OF MUSIC Elizabeth Hellmuth Margulis
PSYCHOPATHY Essi Viding
PSYCHOTHERAPY Tom Burns and Eva Burns-Lundgren
PUBLIC ADMINISTRATION Stella Z. Theodoulou and Ravi K. Roy
PUBLIC HEALTH Virginia Berridge
PURITANISM Francis J. Bremer
THE QUAKERS Pink Dandelion
QUANTUM THEORY John Polkinghorne
RACISM Ali Rattansi
RADIOACTIVITY Claudio Tuniz
RASTAFARI Ennis B. Edmonds
READING Belinda Jack
THE REAGAN REVOLUTION Gil Troy
REALITY Jan Westerhoff
THE REFORMATION Peter Marshall
RELATIVITY Russell Stannard

RELIGION IN AMERICA Timothy Beal
THE RENAISSANCE Jerry Brotton
RENAISSANCE ART Geraldine A. Johnson
REPTILES T. S. Kemp
REVOLUTIONS Jack A. Goldstone
RHETORIC Richard Toye
RISK Baruch Fischhoff and John Kadvany
RITUAL Barry Stephenson
RIVERS Nick Middleton
ROBOTICS Alan Winfield
ROCKS Jan Zalasiewicz
ROMAN BRITAIN Peter Salway
THE ROMAN EMPIRE Christopher Kelly
THE ROMAN REPUBLIC David M. Gwynn
ROMANTICISM Michael Ferber
ROUSSEAU Robert Wokler
RUSSELL A. C. Grayling
RUSSIAN HISTORY Geoffrey Hosking
RUSSIAN LITERATURE Catriona Kelly
THE RUSSIAN REVOLUTION S. A. Smith
THE SAINTS Simon Yarrow
SAVANNAS Peter A. Furley
SCEPTICISM Duncan Pritchard
SCHIZOPHRENIA Chris Frith and Eve Johnstone
SCHOPENHAUER Christopher Janaway
SCIENCE AND RELIGION Thomas Dixon
SCIENCE FICTION David Seed
THE SCIENTIFIC REVOLUTION Lawrence M. Principe
SCOTLAND Rab Houston
SECULARISM Andrew Copson
SEXUAL SELECTION Marlene Zuk and Leigh W. Simmons
SEXUALITY Véronique Mottier
SHAKESPEARE'S COMEDIES Bart van Es
SHAKESPEARE'S SONNETS AND POEMS Jonathan F. S. Post
SHAKESPEARE'S TRAGEDIES Stanley Wells
SIKHISM Eleanor Nesbitt
THE SILK ROAD James A. Millward
SLANG Jonathon Green
SLEEP Steven W. Lockley and Russell G. Foster
SOCIAL AND CULTURAL ANTHROPOLOGY John Monaghan and Peter Just
SOCIAL PSYCHOLOGY Richard J. Crisp
SOCIAL WORK Sally Holland and Jonathan Scourfield
SOCIALISM Michael Newman
SOCIOLINGUISTICS John Edwards
SOCIOLOGY Steve Bruce
SOCRATES C. C. W. Taylor
SOUND Mike Goldsmith
SOUTHEAST ASIA James R. Rush
THE SOVIET UNION Stephen Lovell
THE SPANISH CIVIL WAR Helen Graham
SPANISH LITERATURE Jo Labanyi
SPINOZA Roger Scruton
SPIRITUALITY Philip Sheldrake
SPORT Mike Cronin
STARS Andrew King
STATISTICS David J. Hand
STEM CELLS Jonathan Slack
STOICISM Brad Inwood
STRUCTURAL ENGINEERING David Blockley
STUART BRITAIN John Morrill
SUPERCONDUCTIVITY Stephen Blundell
SYMMETRY Ian Stewart
SYNAESTHESIA Julia Simner
SYNTHETIC BIOLOGY Jamie A. Davies
TAXATION Stephen Smith
TEETH Peter S. Ungar
TELESCOPES Geoff Cottrell
TERRORISM Charles Townshend
THEATRE Marvin Carlson
THEOLOGY David F. Ford
THINKING AND REASONING Jonathan St B. T. Evans
THOMAS AQUINAS Fergus Kerr
THOUGHT Tim Bayne
TIBETAN BUDDHISM Matthew T. Kapstein
TIDES David George Bowers and Emyr Martyn Roberts
TOCQUEVILLE Harvey C. Mansfield
TOPOLOGY Richard Earl
TRAGEDY Adrian Poole

TRANSLATION Matthew Reynolds
THE TREATY OF VERSAILLES
Michael S. Neiberg
THE TROJAN WAR Eric H. Cline
TRUST Katherine Hawley
THE TUDORS John Guy
TWENTIETH-CENTURY BRITAIN
Kenneth O. Morgan
TYPOGRAPHY Paul Luna
THE UNITED NATIONS
Jussi M. Hanhimäki
UNIVERSITIES AND COLLEGES
David Palfreyman and Paul Temple
THE U.S. CONGRESS Donald A. Ritchie
THE U.S. CONSTITUTION
David J. Bodenhamer
THE U.S. SUPREME COURT
Linda Greenhouse
UTILITARIANISM
Katarzyna de Lazari-Radek
and Peter Singer
UTOPIANISM Lyman Tower Sargent
VETERINARY SCIENCE James Yeates
THE VIKINGS Julian D. Richards
VIRUSES Dorothy H. Crawford
VOLTAIRE Nicholas Cronk
WAR AND TECHNOLOGY
Alex Roland
WATER John Finney
WAVES Mike Goldsmith
WEATHER Storm Dunlop
THE WELFARE STATE David Garland
WILLIAM SHAKESPEARE
Stanley Wells
WITCHCRAFT Malcolm Gaskill
WITTGENSTEIN A. C. Grayling
WORK Stephen Fineman
WORLD MUSIC Philip Bohlman
THE WORLD TRADE
ORGANIZATION Amrita Narlikar
WORLD WAR II Gerhard L. Weinberg
WRITING AND SCRIPT
Andrew Robinson
ZIONISM Michael Stanislawski

Available soon:

KOREA Michael J. Seth
TRIGONOMETRY
Glen Van Brummelen
THE SUN Philip Judge
AERIAL WARFARE Frank Ledwidge
RENEWABLE ENERGY Nick Jelley

For more information visit our website

www.oup.com/vsi/

Richard Earl

TOPOLOGY

A Very Short Introduction

OXFORD
UNIVERSITY PRESS

Great Clarendon Street, Oxford, OX2 6DP
United Kingdom

Oxford University Press is a department of the University of Oxford. It furthers the University's objective of excellence in research, scholarship, and education by publishing worldwide. Oxford is a registered trade mark of Oxford University Press in the UK and in certain other countries

First edition published in 2019

Published in the United States of America by Oxford University Press
198 Madison Avenue, New York, NY 10016, United States of America

British Library Cataloguing in Publication Data
Data available

Library of Congress Control Number: 2019949429

ISBN 978–0–19–883268–3

Printed and bound by
CPI Group (UK) Ltd, Croydon, CR0 4YY

In memory of
Dan Lunn.
Friend, colleague, tutor.

Contents

Acknowledgements

Thanks go to Martin Galpin, Andy Krasun, Natalie Lane, Marc Lackenby, Kevin McGerty for their comments on and help with draft chapters. Thanks especially to Marc for his encouragement to write the book.

List of illustrations

Chapter 1
What is topology?

As you read this, passengers the world over are travelling on metro (or subway or underground) trains. There are around 60 billion individual journeys made annually on such metro systems. But whether this be in Tokyo, London, São Paolo, New York, Shanghai, Paris, Cairo, Moscow, those travellers are perusing maps for their journeys that are crucially different from maps in atlases or seen on geography classroom walls. Foremost in the minds of those passengers are the connections they need to make—getting out at the right station and changing to the correct new line. They are not interested in whether the map's left–right lines do indeed run west–east, or whether they really did make a right angle turn when they changed lines, as depicted on the metro map.

The oldest metro network in the world is the London Underground. When first produced, the underground maps superimposed the different train lines onto an actual (geographically accurate) map of London, as shown in Figure 1(a). A first version of the current map was designed by Harry Beck in 1931 as in Figure 1(b). Beck's map, and the current underground map, are not wrong. Rather they transparently show information important to travellers—for example, the various connections between lines and the number of stops between stations. It is an early example of a *topological* map and demonstrates the different focus of topology—which is all about shape, connection, relative position—compared with that

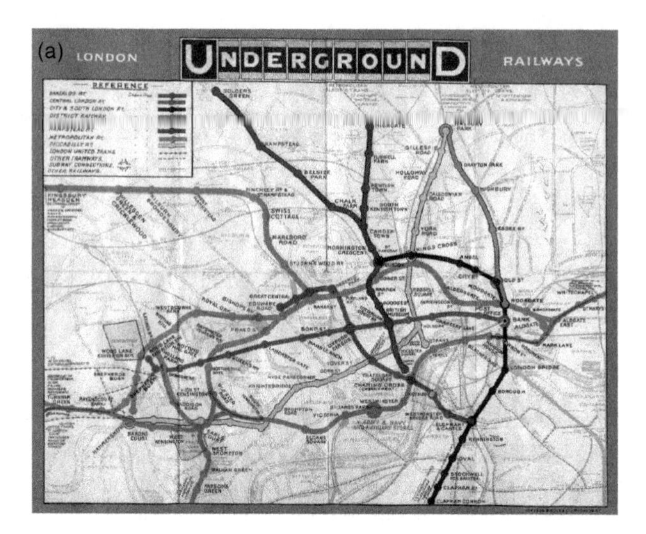

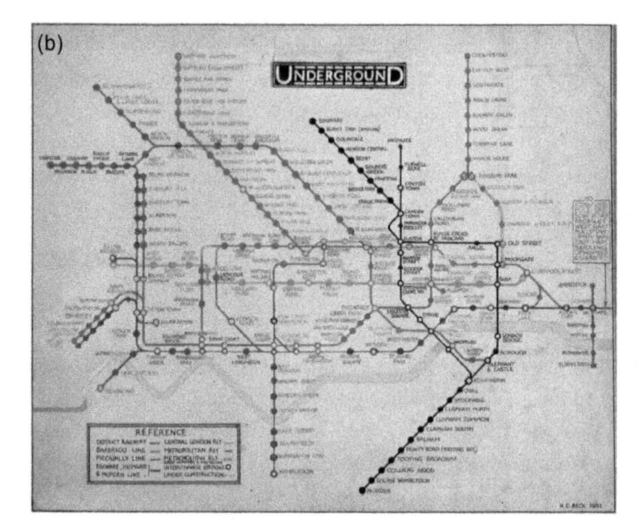

1. London underground maps (a) Geographically accurate 1908 map, (b) Beck's topological 1931 map.

of geometry (or geography) which is about more rigid notions such as distance, angle, and area.

Topology is now a major area of modern mathematics, so you may be surprised to learn that an appreciation of topology came late in the history of mathematics. The word topology—meaning 'the study of place'—wasn't even coined until 1836. ('Geometry' by comparison is an ancient Greek word and 'algebra' is an Arabic word, with its mathematical meaning dating back to the 9th century.) Just why this was the case is not a simple question to address, though we will see some aspects of topology developed as mathematicians sought to put their subject on a more rigorous footing. Topology is a highly visual subject that lends itself to an informal treatment and this book will give you a sense of topology's ideas and its technical vocabulary.

A topologist's alphabet

As a first example, to convey how differently topologists and geometers see objects, consider what capital letters a topologist would deem to be the 'same'. Using the sans serif font, the four letters

E F T Y

are all topologically the same. They are not *congruent*, meaning that none of the letters can be picked up, and rotated or reflected, and then put down as one of the other letters. But I hope you can envisage, if allowed to bend, stretch, or shrink the letters, how any of them might be transformed into one of the others.

To a topologist these four letters are **homeomorphic** to one another. The geometer would notice that the angle made by the arms of the Y is different from any angle found in the other letters. The topologist, on the other hand, would be happy to flatten the arms of the Y, and stretch its body a little, to give the T shape. Likewise the E could have its bottom rung bent around to the vertical, and then shortened somewhat, to make the F. Finally doing the same to the top of the F would make a T on its side. These four letters can be continuously deformed into one another and back again. Broadly speaking this is what it is to be *homeomorphic*, to be topologically the same.

But what is it about these letters that makes them topologically different from other letters? Another collection of letters that are topologically the same as one another is:

C G I J L M N S U V W Z.

Topologically all of these are equivalent to a line segment and it's not hard to imagine how each might be formed by bending and stretching a suitably mutable letter I. So hopefully you're convinced that the letters in the second list are all homeomorphic to one another, but what makes this second collection topologically different from the first list?

Note, for each letter E, F, T, Y, that every point lies on a distorted bit of line with one exception. In each of these letters there is a single point that might be described as a *T-junction*. These T-junctions are highlighted below.

E F T Y

One way in which these T-junctions are special is that, if removed, the remainder of the letter is disconnected into three parts; the removal of any other point would leave just two parts remaining. In whatever ways we might bend and deform an E the deformed version would still include a single T-junction. As none of the second set C...Z has such a T-junction then none of them can be a deformed version of an E (or an F, T, or Y).

This gives a genuine sense of how mathematicians resolve the question: *are two shapes the same topologically?* This either amounts to finding some means of continuously deforming one into the other, or involves finding some **topological invariant** of one that does not apply to the other. The word *invariant* is used in different contexts in mathematics: for example, if you shuffle a pack of cards, there will still remain fifty-two cards afterwards and four suits, these are invariants; but the top card may have changed and the jack of clubs may no longer come before the eight of diamonds, and so such facts aren't invariants of a shuffle. A *topological invariant* is something immutable about a shape, no matter how we stretch and deform it. In the above example we used the presence of a T-junction as our topological invariant. You might note that an E includes four right angles whilst an F contains only three. The presence of four right angles is a geometric invariant and so shows that E and F are not congruent (i.e. not geometrically the same), but—working topologically—we are permitted to unbend these right angles and so right angles are not important from a topological point of view. Rather they're mutable aspects of a shape and not topological invariants.

The remaining twenty-six letters, grouped topologically, break down as:

DO, KX, AR, B, PQ, H.

You might want to take a moment thinking about what makes an A different from a P or O different from Q. In fact, the O introduces an important topological invariant that separates it from both I and E. The shape of the O is different as it makes a loop. Technically O is not **simply connected**, a topic we will discuss more in Chapter 5.

Euler's formula

One of the first topological results was due to Leonhard Euler (pronounced 'oil-er'), a titan of 18th-century mathematics and one of the most prolific mathematicians ever; his formula dates to around 1750. The result relates—at first glance—to **polyhedra**, three-dimensional objects such as cubes and pyramids (Figure 2). It is also so fundamental—a straightforward observation at least—that it is surprising ancient Greek mathematicians missed it.

Looking at the cube, we can see that it is made up of **vertices** (the corners of the cube—the singular is 'vertex'), these vertices being connected by **edges** and that these edges then bound (square) **faces**. For the cube the number of vertices V equals 8, there are $E = 12$ edges and $F = 6$ faces. For the (square-based) pyramid we have $V = 5$, $E = 8$, and $F = 5$. No pattern may be evident immediately but if we include the four other so-called *Platonic solids*—tetrahedron, octahedron, dodecahedron, icosahedron (Figure 4)—and other familiar polyhedra, we create Table 1.

(We shall see soon that the truncated icosahedron is familiar to us—just not by that name!)

For all the geometry that the ancient Greeks knew, it seems striking that this pattern eluded them, but we will prove now—or more honestly sketch a proof of—**Euler's formula** which states, for a polyhedron with V vertices, E edges, and F faces, that

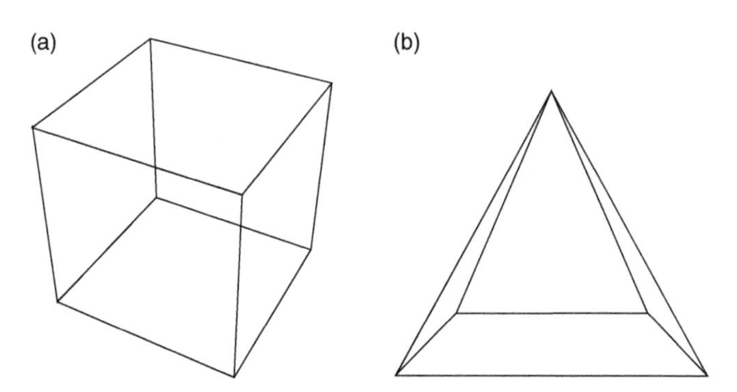

2. Examples of polyhedra (a) A cube, (b) A Square-based pyramid.

Table 1. Vertices, edges, faces for various polyhedra

Polyhedron	V	E	F	$V-E+F$
Tetrahedron	4	6	4	2
Pyramid	5	8	5	2
Cube	8	12	6	2
Octahedron	6	12	8	2
Dodecahedron	20	30	12	2
Icosahedron	12	30	20	2
Truncated Icosahedron	60	90	32	2

$$V-E+F=2.$$

In the proof our aim will be to begin with a polyhedron and manipulate it in certain ways—for example, we might remove or subdivide faces—but in all cases we will carefully track the effect our manipulation has (if any) on the number $V-E+F$. If, after such manipulations, we arrive at a simplified situation where we know what $V-E+F$ equals, and we know the effects our manipulations had on that number, then we may be able to work backwards to find what $V-E+F$ was originally.

We begin then with a polyhedron, and first remove one of the faces. This has the effect of reducing $V-E+F$ by 1 as F has decreased by 1. Now that the polyhedron has a missing face—effectively the polyhedron has been punctured—it can be flattened into the plane, taking care that all the vertices, edges, faces present on the punctured polyhedron remain and are connected in the plane in the same manner they were on the punctured polyhedron. For example, if we removed one face from a cube and flattened the remaining cube then we would have something like Figure 3(a).

The next manipulation is to subdivide each of the flattened faces into triangles—as has been done to the flattened cube in Figure 3(b).

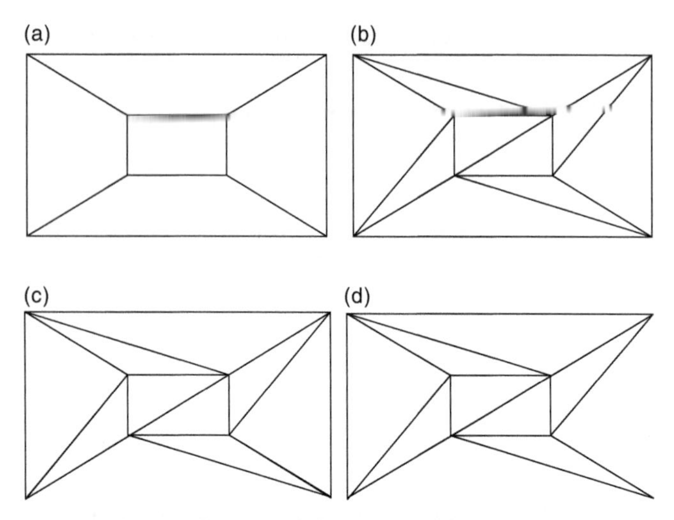

3. Manipulations of a cube to find its Euler number (a) A flattened, punctured cube, (b) With flattened faces triangulated, (c) Removing a triangle, (d) Removing a triangle.

Introducing a single triangle has the effect of increasing F by 1—what was one face becomes split into two—of increasing E by 1—the new edge, introduced to make a triangle—and doesn't change V. So there is no overall effect to $V - E + F$ as we keep introducing triangles; the increase of 1 to F, a term that is added in the formula, is precisely balanced by the increase in E which is a term we subtract. When this has been done for each flattened face (as in Figure 3(b)) then $V - E + F$ is still just one less than it was originally.

We now remove the triangles one at a time. For example, if we remove the bottom triangle from Figure 3(b) to make Figure 3(c), then we remove one edge and one face and, by the same reasoning as before, this has no overall effect on $V - E + F$.

Similarly, we might then remove the right-most triangle to create Figure 3(d), the manipulation again having no effect on $V - E + F$. But the bottom triangle of Figure 3(d) is connected differently.

If we remove that triangle then we remove 2 edges, 1 face, and 1 vertex. The algebra is a little more complicated this time, but again removing 1 vertex and 1 face means $V - E + F$ goes down 2 but this is countered by removing 2 edges as E is a term we subtract. Or if you prefer more formal algebraic reasoning, we are just saying

$$(V-1)-(E-2)+(F-1)=V-E+F.$$

Let's summarize what's happened so far:

- We removed a face and $V - E + F$ decreased by 1.
- We flattened the polyhedron into the plane—all V, E, F remain, so no change to $V - E + F$.
- We subdivided the flattened faces into triangles—this had no effect on $V - E + F$.
- We kept removing triangles from the edge of the flattened polygon—each removal having no effect on $V - E + F$.

Eventually only a single triangle will remain, having removed all others. A triangle has a single face, three vertices, and three edges, so that $V - E + F$ equals $3 - 3 + 1 = 1$. This is the value of $V - E + F$ that we finish with. The only manipulation that ever changed $V - E + F$ was that very first removal of a face which reduced it by one; initially then it was the case that

$$V-E+F=2.$$

This 'sketch proof' was given by Augustin-Louis Cauchy in 1811. It's worth highlighting there are several 'i's still to be dotted to make a proof with which a professional mathematician would be happy, but also noting how much of the idea of the proof is genuinely here. We didn't take care describing how we removed triangles from the boundary of the flattened polygon; if we'd been careless we might have removed a triangle that disconnected the flattened polygon into two separate polygons, and we should have taken time to make sure such an occasion can always be avoided. Other issues will become more apparent in Chapter [illegible] but these 'I's can indeed be dotted. In *Proofs and Refutations*, the Hungarian philosopher Imre Lakatos used the specific example of Euler's formula, and historical efforts to prove it, to highlight how hard it can sometimes be to generate a watertight proof and to also raise the question of when a theorem properly becomes part of mathematics or has mathematical content.

It's also worth mentioning that René Descartes had, over a century before Euler, demonstrated a theorem for polyhedra that is equivalent to Euler's formula; his theorem was in terms of 'angular defects' at vertices. All of a sudden we are back in the geometrical world and it's less than clear that there is a genuinely new subject, an importantly different way of mathematical thinking, that Euler's fingertips were brushing against. Euler's formulation encourages appreciation of the result as something a little new—in Euler's terminology, rather than Descartes's, it's much clearer that the connection of the vertices, edges, and faces is what counts, but historically we are still a long time from a deeper appreciation of topology as a fundamental mode of mathematical thinking.

There are five Platonic solids

Platonic solids are polyhedra with regular faces that are all congruent (geometrically the same) and which meet in the same manner at each vertex. There are infinitely many regular polygons—equilateral triangles, squares, regular pentagons, etc.—but in 3D it turns out that there are just five regular solids which have been known since antiquity. These are shown in Figure 4 and with Euler's formula we can show there are just these five.

Consider a regular polyhedron with V vertices, E edges, and F faces. As the solid is regular then each face is bounded by the same number of edges; let's call this number n. Likewise there is a

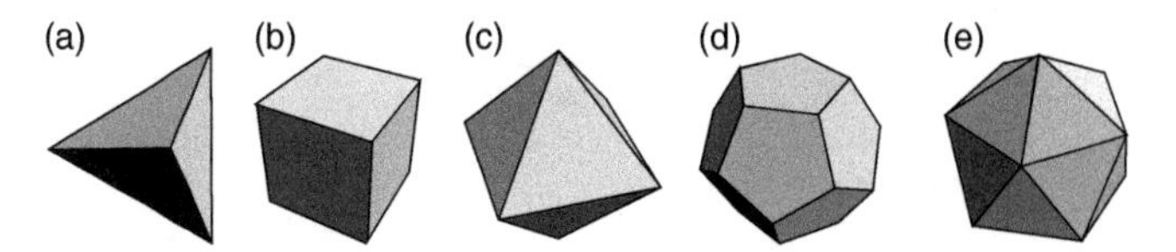

4. The Platonic solids (a) Tetrahedron, (b) Cube, (c) Octahedron, (d) Dodecahedron, (e) Icosahedron.

common number m for how many edges meet at each vertex. So, with the cube, $n = 4$ (the faces are squares) and $m = 3$ (three edges meet at each vertex). Continuing with the cube as our example for now, think about how we can make a cube by gluing together the edges of six squares.

We begin with 6 separate squares so that, before any gluing happens, there are 6 squares, 24 edges, and 24 vertices. Note that to make the cube it takes two 'unglued' edges to make each edge of the cube (which agrees with there being $24/2 = 12$ edges) and it takes three 'unglued' vertices to make a single vertex of the cube (again there are $24/3 = 8$ vertices). More generally, when we have F faces each with n edges, we would have nF edges before any gluing. It takes two of these unglued edges to make a single edge of the polyhedron which then has $E = nF/2$ edges. There are as many unglued vertices as unglued edges, namely $V = 2E$, and these will be glued together to make $V = 2E/m$ vertices on the polyhedron as it takes m unglued vertices to make one glued vertex on the solid.

Putting these expressions for V and F into Euler's formula we get

$$2 = V - E + F = \frac{2E}{m} - E + \frac{2E}{n}.$$

(The equation $E = nF/2$ has been rearranged to make F the subject of the equation so that $F = 2E/n$.) We can then divide both sides of the above equation by $2E$ and rearrange to find

$$\frac{1}{2}+\frac{1}{E}=\frac{1}{m}+\frac{1}{n}.$$

As $1/E$ is positive this means that

$$\frac{1}{m}+\frac{1}{n}>\frac{1}{2}.$$

So m and n can't both be very large as then $1/m$ and $1/n$ would be very small and their sum would not exceed $1/2$. Also recall m and n are positive whole numbers, so there aren't many options and it's not hard to find all their possible values.

It's impossible for *both* m and n to exceed 4 as then $1/m + 1/n$ would be less than $\frac{1}{4} + \frac{1}{4} = \frac{1}{2}$. So either $m = 3$ or $m = 4$ or $n = 3$ or $n = 4$ (with perhaps more than one of these being true). For example, if $m = 3$, the only n for which the inequality is true are $n = 3, 4, 5$. If $n \geqslant 6$ then $1/m + 1/n$ equals $1/3 + 1/6 = 1/2$ or less and the inequality is not true. For the three cases of $m = 3$ and $n = 3, 4, 5$ we have

$$\frac{1}{E}=\frac{1}{3}+\frac{1}{3}-\frac{1}{2}=\frac{1}{6} \text{ giving } E=6.$$

$$\frac{1}{E}=\frac{1}{3}+\frac{1}{4}-\frac{1}{2}=\frac{1}{12} \text{ giving } E=12.$$

$$\frac{1}{E}=\frac{1}{3}+\frac{1}{5}-\frac{1}{2}=\frac{1}{30} \text{ giving } E=30.$$

A similar calculation for $m = 4$ leads to $n = 3$, $E = 12$, and when $m = 5$ we find $n = 3$, $E = 30$. In all these five cases we can use the previous formulas to work out the numbers of vertices $V = 2E/m$ and faces $F = 2E/n$. We can put the full details into Table 2.

If we are seeking to be rigorous here, we should really point out that the previous calculations show that there are *at most* five possible pairs of values that m, n can take. Those calculations limit the possibilities, but do not necessarily mean that there is a Platonic solid for each of these cases, nor preclude there being

Table 2. Possible m and n values for the Platonic solids

m	n	V	E	F	Platonic solid
3	3	4	6	4	Tetrahedron
3	4	8	12	6	Cube
4	3	6	12	8	Octahedron
3	5	20	30	12	Dodecahedron
5	3	12	30	20	Icosahedron

more than one Platonic solid for permitted m and n—it might be that there are two different Platonic solids with three pentagons meeting at each vertex. Listed in Table 2 are the five Platonic solids, and so we can see that there is at least one solid for permitted m, n. And it's not hard to appreciate why there can be at most one. In the case where $m = 3$, $n = 4$ then three squares meet at each vertex; seemingly this only tells us something about parts of the solid, but if we follow this recipe of attaching three squares at each vertex then there is only one way to progress building up the solid—it's not clear that this recipe will actually lead to a complete solid, but it does show that there can be at most one Platonic solid for each allowed m, n.

(As an aside, you may have noticed that $m = n = 4$ and $m = 3$, $n = 6$ and $m = 6$, $n = 3$ are solutions if we permit E to be infinite. These 'solutions' correspond to tessellations of the plane where four squares meet at a vertex, where three regular hexagons meet at a vertex (as with honeycombs), and where six equilateral triangles meet at a vertex. There are also some patterns apparent in Table 2 for the values of V, E, F for the cube and octahedron, and likewise the dodecahedron and icosahedron. This is because these solids are *dual* to one another—this means that the midpoints of the faces of a cube make an octahedron, and vice versa; likewise the dodecahedron is dual to the icosahedron and the tetrahedron is self-dual in this sense.)

Footballs

In 2017 the mathematics popularizer Matt Parker began a petition seeking to get road signs to football stadia corrected.

You may not have noticed the inaccuracy of such signs in the past, perhaps being happy just to know you're travelling the right way for the game. But it's clear (Figure 5) that the sign's football does not resemble the actual football that Matt is carrying. A football's surface is made from pentagons and hexagons and the everyday football is more formally known as a **truncated icosahedron**. (It can be created from an icosahedron by planing down, around each vertex, the five edges meeting there. If we plane down one-third of each of those five edges, we create a new pentagonal face and continued planing eventually shrinks all the triangular faces to hexagons.) I think though the irksome principle for Matt was not that the sign's football was badly drawn, it was in fact *impossibly* drawn.

There is no way that a sphere can be made by stitching together hexagons as shown on the sign. That would be an example where

5. Matt Parker and his football.

$m = 3$ and $n = 6$, using the previous notation, and whilst we can cover the plane in this way—which may be why the sign looks plausible at first glance—making a football this way is mathematically impossible.

In fact, Euler's formula shows us how many pentagons and hexagons there are on a football. Recalling how to truncate an icosahedron we see there are as many pentagons as original vertices (12) and hexagons as original faces (20), but Euler's formula can show this is the *only* way to construct such a football. Say a football has P pentagonal faces and H hexagonal faces. Then, before gluing these together, we have $5P + 6H$ unglued edges and the same number of unglued vertices. Looking at Matt's ball we can see that (i) two unglued edges are needed to make an edge on the football and (ii) three unglued vertices make a vertex with (iii) two hexagons and one pentagon meeting at a vertex; from (iii) we see there are twice as many 'unglued' vertices collectively on the hexagons as on the pentagons. So

$$F = P + H, \quad E = (5P+6H)/2, \quad V = (5P+6H)/3, \quad 2 \times 5P = 6H.$$

If we put these values into Euler's formula $V - E + F = 2$ we find that

$$(5P+6H)/3 - (5P+6H)/2 + P + H = 2$$

which simplifies and rearranges to $P = 12$ and the equation $10P = 6H$ yields $H = 20$. This, then, is the only way to make a football if we follow the rules (i), (ii), (iii).

Graph theory

Let's change tack a little and consider the following problem. The square PQRS in Figure 6 has diagonally opposite vertices P and R, Q and S. If we were to draw curves from P to R, and from Q to S, curves which remain within the square as in Figure 6, then surely those curves would have to cross at

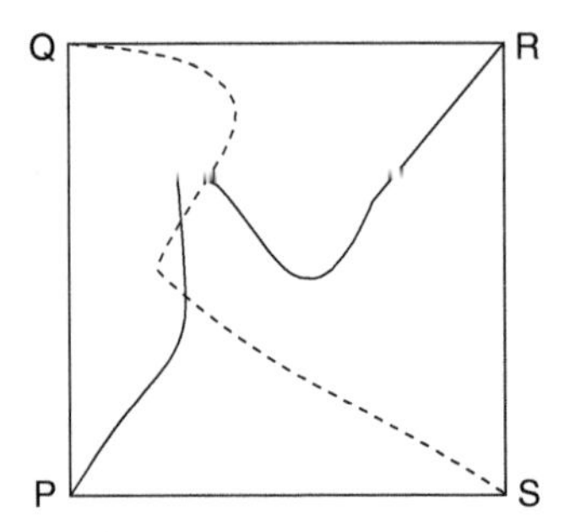

6. Diagonals in a square must intersect Curves *PR* and *QS* in the square *PQRS*.

least once. (In Figure 6 there are three intersections.) This seems obvious—and is true—but how would you go about proving this?

Before more is said, it might be worth stressing how characteristic of a topological question this is. The curves PR and QS need to connect their end points. Those curves don't need to be polygonal, or have well-defined gradients, or be defined by specific functions. They need to connect the end points in some *continuous* sense—fuller details in Chapter 3—but they are otherwise general paths from P to R and from Q to S that remain in the square.

At first glance, this problem might seem quite removed from the polyhedra we were just discussing. However, Figure 6 doesn't look that different from Figures 3(a)–(d). We have vertices (P, Q, R, S and any points where the curves PR and QS meet), edges running between these vertices (though admittedly they're now curved), and we have faces bounded by those edges. It was crucial to the proof of Euler's formula that $V - E + F = 1$ for each of Figures 3(a)–(d). If we also include the outside region as a face—essentially the one removed so we could flatten the polyhedron—then we arrive back at Euler's formula $V - E + F = 2$. (By this reckoning $V = 7$, $E = 12$, $F = 7$ in Figure 6.)

So suppose, somehow, we could draw curves PR and QS in the square PQRS which *don't* intersect. We'd then find

- $V = 4$ the four corners P, Q, R, S.
- $E = 6$ the square's four sides and the curves PR, QS.
- $F = 5$ the outside of the square, above PS, below QR, right of PQ, left of SR.

But this leaves us with $V - E + F = 3 \neq 2$ and so such a scenario is impossible by Euler's formula.

Graph theory is an area of mathematics that models networks in a wide sense: physical, biological, and social systems, variously representing transport networks, computer networks, website structure, evolution of words across languages and time in philology, migrations in biology, etc. A *graph* is a collection of points called vertices, with these vertices connected by edges. We will also assume that graphs are *connected*, meaning that there is a walk between any two vertices along the edges. This definition may be extended to include one-way edges—*directed graphs* or *digraphs*—and *weights* might be introduced to edges representing the difficulty—in terms of time, distance, or cost—of travelling along a particular edge.

Some graphs are *planar*, meaning that they can be drawn in the plane without their edges crossing (at points that aren't vertices). The two graphs K_5 and $K_{3,3}$ in Figure 7 are importantly *not* planar. The *complete graph* on 5 vertices, denoted K_5, has a single edge between each pair of the 5 vertices making 10 edges. You might think that K_5 is planar as it's drawn in Figure 7(a)—the point is that, so drawn, many of the edges' crossings don't occur at vertices and to deem these crossings as vertices would mean we were no longer considering K_5 which has *only* 5 vertices. If we could properly draw K_5 in the plane there would be 10 triangular faces $v_1v_2v_3$, $v_1v_2v_4$, through to $v_3v_4v_5$. We'd then have that $V - E + F$ equals $5 - 10 + 10 = 5 \neq 2$ and so K_5 is not planar.

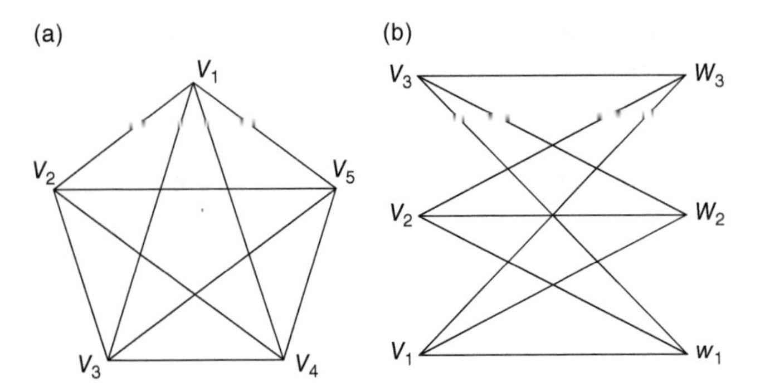

7. Two non-planar graphs (a) The complete graph K_5, (b) The complete bipartite graph $K_{3,3}$.

$K_{3,3}$ is the *complete bipartite graph* between two trios of vertices. A somewhat subtler argument shows $K_{3,3}$ is not planar. Note that $K_{3,3}$ has $V = 6$ vertices and $E = 9$ edges. If drawn in the plane this would mean $F = 2 - V + E = 5$. But a face of $K_{3,3}$ would have at least four edges as its perimeter necessarily runs from a v to a w to a different v and to a w and only then may return to the original v. So, counting the edges by going around all the faces, we would get a total of at least $5 \times 4 = 20$ edges. However, as an edge can bound at most two faces this would mean we'd have at least $20/2 = 10$ edges, which is our required contradiction.

The Polish mathematician Kazimierz Kuratowski proved in 1930 that a graph is planar precisely when neither a copy of K_5 nor $K_{3,3}$ can be found within the graph. We will in due course see that $K_{3,3}$ can be drawn on other surfaces such as a torus (Figure 13(c)).

Nasty surprises

Euler arrived at his formula a century before the word topology was coined. His formula is characteristic of a visual side of topology naturally aligned with geometry. But topology, as a subject, would develop along various themes and in particular had an important role in the foundational work mathematicians were doing around the start of the 20th century. As I hinted earlier, topology's rise may have been hampered by a traditional mindset that some of its questions had obvious answers. For example Camille Jordan, as late as the 19th century, *proved* the following: a curve in the plane, which does not cross itself and which finishes back where it began—a curve which we would now call a **Jordan curve**—splits the plane into two regions, the technical phrase for these regions being *connected components*. One of these regions is bounded, the *inside*, and the other is unbounded, the *outside*, and this is the **Jordan curve theorem**. Earlier mathematicians would have happily thought this obvious and the first rigorous proof didn't appear until 1887. You may agree with those earlier mathematicians that the result can be safely assumed. Maybe even the Pollock-like Jordan curve in Figure 8(a) does not sway your view of the intuitiveness of the result.

In Figure 8(b) is the *Knopp–Osgood curve* which, for all its fractal-like appearance, is a Jordan curve. Astonishingly it has a positive area—that is the curve *itself* has positive area, we're not referring to some region that it bounds. Would you have said a moment ago that it's obvious that curves can't themselves have area?

You shouldn't worry too much in the sense that most things those early mathematicians thought to be true turned out to be true, once properly understood and qualified, but mathematicians towards the end of the 19th century were getting nervous about the rules and assumptions that mathematics relied on.

A related problem within topology at that time was rigorously defining what *dimension* means. Again this had previously been treated as an intuitive concept, only for mathematicians to begin finding space-filling curves that pass through every point in the

8. Complicated Jordan curves (a) A more complicated Jordan curve, (b) A Jordan curve with positive area.

plane or other weird-and-wonderful spaces that can reasonably be assigned dimensions that are not whole numbers—spaces that would now be called **fractals**.

An early theme of topology was this **general topology** or **point-set topology** seeking to address what it means to be a set, to be a space, etc. *Metric* and *topological spaces* were introduced—to be discussed in Chapter 4—each being attempts to describe general structures where continuity could be defined. Set theory deals with collections that are essentially just things-in-a-bag. This general topology sought to define ways in which objects might be considered 'close' to one another, with the aim being to define continuity in a broad setting.

A Flatland mindset

The novella *Flatland: A Romance of Many Dimensions*, written in 1884 by Edwin Abbott, is a satire on Victorian mores. The narrator is 'A Square', an inhabitant of Flatland, a planar world having just two dimensions. The culture of Flatland and the logistics of living in two dimensions are fully described, implicitly highlighting some of the narrow-mindedness of Victorian culture—for example, women are one- rather than two-dimensional beings. The story doesn't explicitly discuss topology, but in its description of worlds with different dimensions and implications for the inhabitants, it provides a useful metaphor for understanding certain aspects of topology.

For example, A Square is visited at one point by A Sphere. Being a three-dimensional object, A Sphere can only be perceived by Flatlanders as a circular cross-section (Figure 9). By moving up and down—relative to Flatland's plane—A Sphere can grow, shrink, and even disappear entirely. In a similar manner, to truly understand the topology of a space, we have to begin thinking like inhabitants of that space.

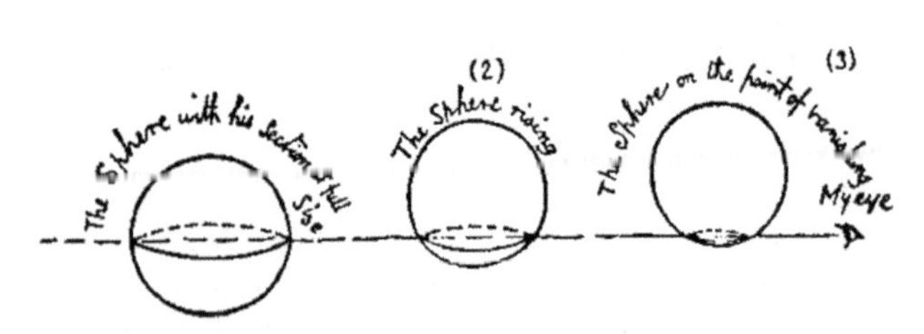

9. Original figure from *Flatland* showing how A Sphere is perceived by A Square.

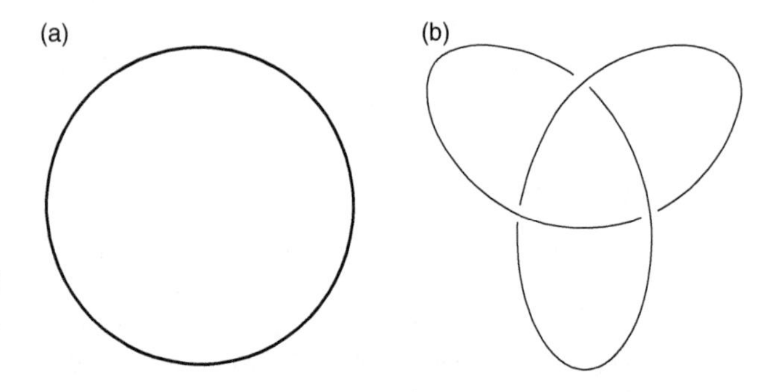

10. The unknot and the trefoil (a) The unknot, (b) The trefoil knot.

Topology is often characterized as rubber-sheet geometry. It's a somewhat clichéd metaphor, but it's also slightly inaccurate. It gives a correct sense of topology being more about shape and less rigid than geometry in its focus. On the other hand, in Chapter 6 we discuss knots, and as a (genuine) knot—like the trefoil—and the (unknotted) circle (Figure 10) cannot be continuously deformed into one another in 3D then you might be tempted to say the circle and trefoil are not homeomorphic, but they are.

The knottedness of the trefoil says something about its position in 3D. In fact, *all* knots are homeomorphic to a circle. To better appreciate this, you might imagine life as an ant living on either the circle or trefoil. As the ant moves around either the unknot or trefoil it has a sense of being on a loop, but the ant has no notion of whether it is living on a knot. It is only by being able to view things from outside the two loops, and looking on from a position in the *ambient space*, that we are able to recognize one loop as knotted as compared with the other. This Flatland mindset will prove useful again later when we meet *subspaces*.

Topology would advance on various fronts in the 19th and 20th centuries. In particular, Bernhard Riemann would early on show the usefulness of a 'topological mindset', introducing *Riemann surfaces* into the study of polynomial equations and demonstrating some deep connections between topology and many other areas of mathematics.

Chapter 2
Making surfaces

The shape of surfaces

Recall Euler's formula states $V - E + F = 2$ for a polyhedron. Various details of the proof were brushed under the carpet, the most significant of these being the claim that, once a face is removed from a polyhedron, the remaining polyhedron can be flattened into the plane. This was true for the polyhedra we were considering, but the claim says something important about the shape of the remaining polyhedron that was perhaps unintentional. In any case, the next example will either make us question what we mean by a polyhedron or have us looking to generalize Euler's formula.

For Figure 11(a)'s 'polyhedron', a count of vertices, edges, and faces shows that $V = 16$, $E = 32$, $F = 16$, giving $V - E + F = 0$ which seems to disprove Euler's formula. We are left with a few alternatives: either the object in Figure 11(a) should not be considered a polyhedron, or we need to restrict Euler's formula to a certain type of polyhedron, or we need to adapt and generalize Euler's formula into a version that remains true for a broader family of polyhedra.

The most obvious issue with this new 'polyhedron' is the hole through its middle. This is not immediately reason enough to exclude it as a polyhedron, but this shape, once a face is removed,

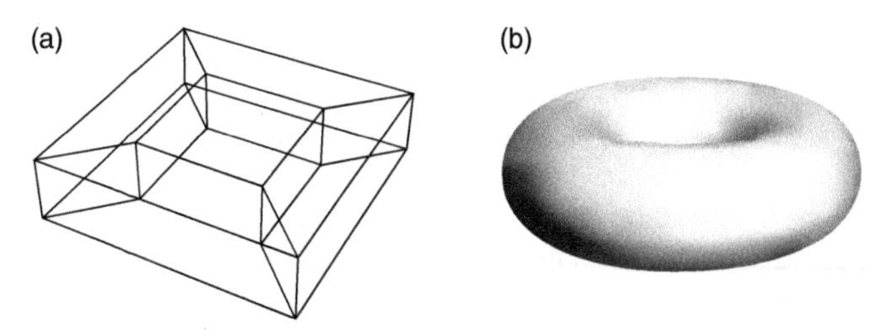

11. The torus (a) A polyhedron with one hole, (b) A torus.

does not leave a remainder that can be flattened into the plane, making our earlier proof invalid. We need either to restrict Euler's formula to polyhedra without holes, or we need to work out the correct $V - E + F$ values for polyhedra with holes.

Recalling the rubbery nature of topology, we might recognize that the polyhedra of Chapter 1 all had the same underlying spherical shape. If allowed to smooth out those polyhedra—the pointy vertices and the ridgy edges—we could transform each of those polyhedra to a sphere, covered with a patchwork of curved faces, just like Matt Parker's football (Figure 5) was covered with curved pentagons and hexagons. But however we smooth down our new polyhedron we can't make a sphere, rather we would make a torus, the shape of a doughnut with a hole through it (Figure 11(b)).

Perhaps then all of the examples of Chapter 1—including the proof of Euler's formula—point to the **Euler number** of the sphere being 2. And Figure 11(a) is a first example suggesting the Euler number of the torus is 0; this would mean the number $V - E + F$ equals 0 however we divide up the torus. All this could become quite involved unless we have a way of efficiently describing surfaces—including more complicated ones than the sphere or torus—and for systematically calculating their Euler numbers, that is the $V - E + F$ value common to all surfaces of a certain underlying shape.

Gluing surfaces together

A useful way of constructing surfaces is to begin with a polygon and pairwise glue together the edges of the polygon, the way a model kit might direct you to 'glue tab A to tab B'. How might we make a torus in this manner? If we begin with a (suitably elastic) square (Figure 12(a)), bend it around (Figure 12(b)), and glue the opposite edges e_1 and e_3 so that the vertices v_1, v_2 get glued respectively to v_4, v_3, and likewise for all other opposite points of e_1 and e_3—as signified by the two arrows—then we will make the cylinder drawn in Figure 12(c). Note that the edges e_2 and e_4 on the original square have become the two circular ends of this cylinder. We can then glue together these circular ends; if we do this so that the opposite points of the original e_2 and e_4 are glued together, then we make a torus as in Figure 13(a).

Note that the four corners of the original square—denoted v_1, v_2, v_3, v_4—have all been glued together to make a single point v on the torus. Similarly, the edges e_1 and e_3 have been glued together to form a circle going around the outside of the torus and e_2 and e_4 have been glued together to form a different circle going through the hole of the torus, these two circles meeting at the point v.

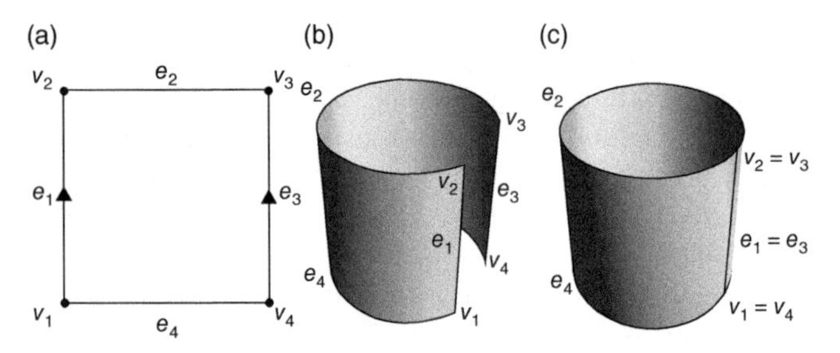

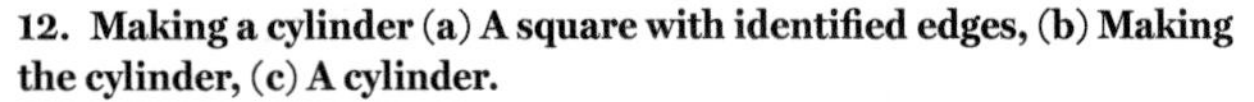
12. Making a cylinder (a) A square with identified edges, (b) Making the cylinder, (c) A cylinder.

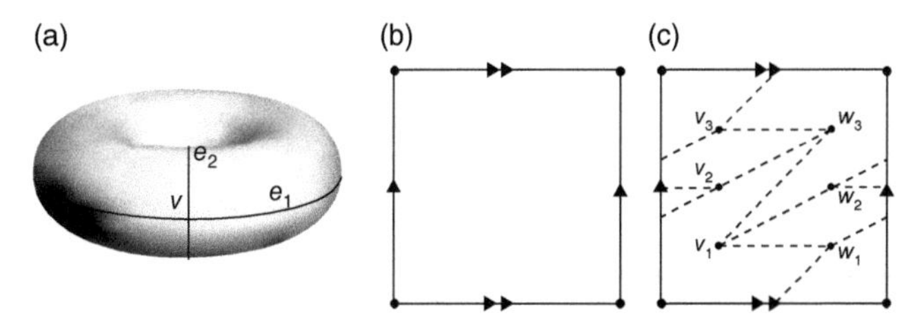

13. Making a torus, (a) Torus with v and e_1, e_2 drawn, (b) Square with gluing instructions, (c) $K_{3,3}$ on the torus.

Importantly, the torus's shape is fully described by a square with directions for how the edges are to be glued together. Mathematicians would draw this square-with-gluing-instructions as shown in Figure 13(b). The single arrows—and importantly the directions in which they're drawn—show how those two edges are glued, and the double arrows (again noting directions) tell us how the other pair of edges is glued.

As an aside, the graph $K_{3,3}$, which we saw can't be drawn in the plane (Figure 7(b)), *can* be drawn on the torus (Figure 13(c)). The reason $K_{3,3}$ can be drawn on the torus is because a torus has a lower Euler number of 0 and, arguing as before, it can then be shown that $F = 3$. If you look carefully at Figure 13(c) you will see that there are indeed three faces (quadrilaterals $v_1w_2v_2w_1$ and $v_3w_3v_2w_2$ and a single hexagonal face $v_1w_2v_3w_1v_2w_3$).

How does all this help with determining Euler numbers? With a single square and gluing directions, we have been able to make an object with the shape of a torus and this is certainly a conciser description of a torus than Figure 11(a) for which $V = 16$, $E = 32$, $F = 16$. But is this glued square enough to work out the Euler number of a torus? The answer is yes if we're careful when thinking about just how the original vertices and edges glue together. On the original square there was just one face—the square's interior—four unglued edges (labelled e_1, e_2, e_3, e_4), and four unglued vertices (labelled v_1, v_2, v_3, v_4). However, once we've followed the gluing directions, those four edges have become the two circular edges on the torus and the four vertices have become one point v, as in Figure 13(a). And there is still one 'face' on the torus—the square's interior has been stretched to become all of the torus except those circles and v. So when we use this glued square to calculate the Euler number of the torus we get the answer

$$V - E + F = 1 - 2 + 1 = 0,$$

which agrees with our calculation from Figure 11(a).

If you prefer Figure 13(a) showing the torus with the four glued vertices becoming one, and the four edges become two loops on the torus, then Figure 13(b) will only appear as an unfinished DIY job. But the single surface obtained from gluing the triangle, pentagon, and square in Figure 14, following all the gluing directions **a, b, . . . f** according to the arrows, might reasonably start stretching your visualization skills. But we can still work out the Euler number of this chimera and seek to understand just what surface we are looking at. This time there are three faces (the triangle, pentagon, and square) and the twelve unglued edges of the polygons make six glued edges **a, b, c, . . . f** on the surface. How many vertices will we ultimately have? There were twelve unglued vertices originally but various of these get glued together as we

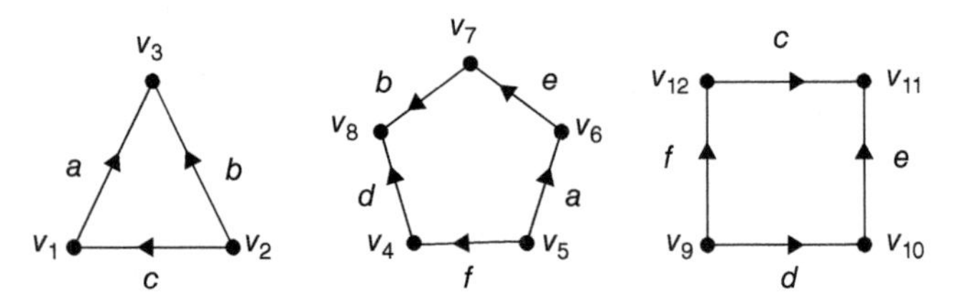

14. Gluing instructions for a triangle, pentagon, and square.

make the surface. For example, v_1 and v_5 are glued together as they are both at the rear end of the edge marked **a**. In fact, we can chase around these gluings to see just how many vertices we have:

v_1 and v_5 are glued (rear end of **a**)
v_5 and v_9 are glued (rear end of **f**)
v_9 and v_4 are glued (rear end of **d**)
v_4 and v_{12} are glued (front end of **f**)
v_{12} and v_2 are glued (rear end of **c**)
v_2 and v_7 are glued (rear end of **b**)
v_7 and v_{11} are glued (front end of **e**)
v_{11} and v_1 are glued (front end of **c**)

So eight different (unglued) vertices $v_1, v_2, v_4, v_5, v_7, v_9, v_{11}, v_{12}$ all get glued together as a single vertex on the surface. In a similar fashion we can see that the remaining four vertices v_3, v_6, v_{10}, v_8 get glued together (in that order). So once made, the surface has 2 vertices, 6 edges, and 3 faces giving an Euler number $V - E + F$ of $2 - 6 + 3 = -1$. Just what surface have we made?

Getting the right answer: subdivisions

We need now to make clear just what surfaces we are considering—**closed** surfaces—and how they can be divided up into vertices, edges, and faces. A closed surface is one without a boundary, such as a torus or sphere, but not the cylinder of Figure 12(c). Our process of making a torus begins with a square and at that point our surface has a boundary consisting of its four edges; when we glue two edges to make a cylinder then the surface still has a boundary, namely its top and bottom circles. Once the torus has been made, no boundary points remain unglued.

Secondly, we can only calculate the correct Euler number of a closed surface if we are careful dividing it up. Cubes, footballs,

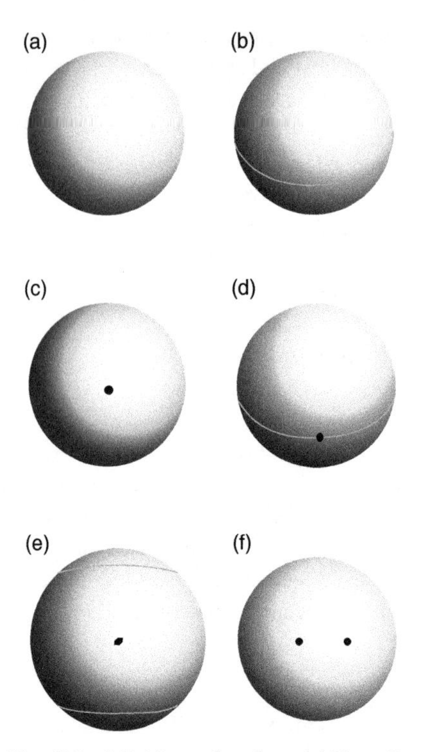

15. Valid and invalid subdivisions of a sphere (**a**) $V = 0, E = 0, F = 1$, (**b**) $V = 0, E = 1, F = 2$, (**c**) $V = 1, E = 0, F = 1$, (**d**) $V = 1, E = 1, F = 2$, (**e**) $V = 1, E = 2, F = 3$, (**f**) $V = 2, E = 0, F = 1$.

dodecahedra, and pyramids are all valid ways of 'subdividing' the sphere into vertices, edges, and faces. In each of these cases we obtained an Euler number $V - E + F$ of 2. However here are other ways we might subdivide the sphere that seemingly produce differing Euler numbers (Figure 15).

In order the values of $V - E + F$ for the six spheres in Figure 15 are 1, 1, 2, 2, 2, 3 and we know the correct Euler number equals 2. So any old subdivision will not lead to a correct calculation of the sphere's Euler number. For a collection of vertices, edges, and faces to make a permissible **subdivision** the following must be true:

- an edge must start and finish in a vertex;
- when two edges meet, they must meet in a vertex;
- faces must be (distorted) polygons.

Looking at these so-called six subdivisions, only two of these are in fact permissible, 15(c) and 15(d). In 15(a), 15(e), 15(f), there is a face that is not a distorted polygon; neither the whole sphere (15(a)) nor the punctured cummerbund (15(e)) nor the twice-punctured sphere (15(f)) are topologically the same as a polygon and so not permissible faces. In 15(b) and 15(e), the edges do not begin and end in a vertex. 15(e) is deliberately given to show that the correct Euler number can be incorrectly calculated.

Looking back at the torus in Figure 13(a) and the surface in Figure 14, we calculated their Euler numbers using subdivisions consistent with the above three rules. Therefore, we correctly calculated their Euler numbers as 0 and –1 respectively.

Connected sums

Given two closed surfaces S_1 and S_2, then we can create their **connected sum** $S_1 \# S_2$. This is a way to glue surfaces together and a useful means of making new surfaces from the few we have so far met. Say S_1 and S_2 each has a subdivision that includes a 'triangular' face bounded by three edges. (The inverted commas here hint that, this being topology, the faces may not be that recognizably triangular in terms of having straight edges.) The connected sum $S_1 \# S_2$ is then created by removing these two triangular faces, so making two holes in the surfaces, and then gluing the two surfaces together along the boundaries of the holes, pairing up the three vertices and three edges with those on the boundary of the second removed face as, for example, in Figure 16.

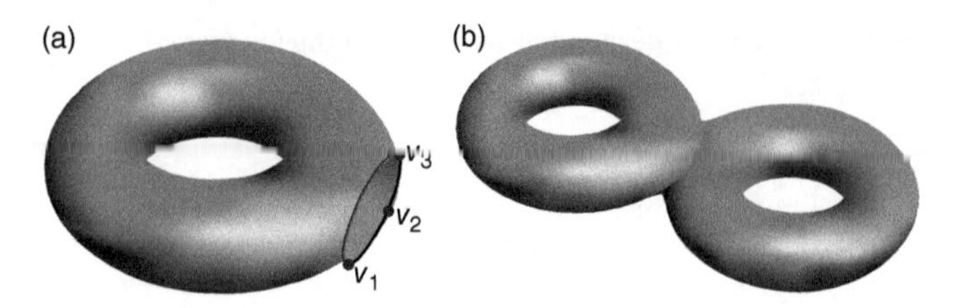

16. Connected sums with tori (a) One torus with a "triangle" missing, (b) A torus with two holes as a connected sum.

Helpfully there is a formula for the Euler number of $S_1\#S_2$. In making the connected sum, we remove two triangular faces, the six different vertices on these triangles are glued to make three vertices on the connected sum, and likewise six edges are glued to make three. So the total number of faces has gone down by 2 and the total numbers of edges and vertices have each gone down by 3. As V and F are added in the formula for the Euler number, and E is subtracted, overall we have

$$\text{Euler number of } S_1\#S_2 = (\text{Euler number of } S_1)+(\text{Euler number of } S_2)-2.$$

Or, if you prefer a more careful algebraic proof, say the original subdivision of S_1 has V_1 vertices, E_1 edges, and F_1 faces and define V_2, E_2, F_2 similarly for S_2. The number of vertices $V_\#$, edges $E_\#$, and faces $F_\#$ on the connected sum is given by

$$V_\# = V_1 + V_2 - 3, \quad E_\# = E_1 + E_2 - 3, \quad F_\# = F_1 + F_2 - 2.$$

Finally

$$\begin{aligned}\text{Euler number of } S_1\#S_2 &= V_\# - E_\# + F_\# \\ &= (V_1 + V_2 - 3) - (E_1 + E_2 - 3) + (F_1 + F_2 - 2) \\ &= (V_1 - E_1 + F_1) + (V_2 - E_2 + F_2) + (-3+3-2) \\ &= (\text{Euler number of } S_1) + (\text{Euler number of } S_2) - 2.\end{aligned}$$

Thinking in terms of connected sums helps us work out the Euler numbers of some more complicated surfaces. We know that a torus $\mathbb{T}$ has an Euler number of 0. The connected sum $\mathbb{T}\#\mathbb{T}$ is a torus with two holes (Figure 16(b)) and we see

$$\text{Euler number of } \mathbb{T}\#\mathbb{T} = 0 + 0 - 2 = -2$$

and similarly the torus with three holes, $\mathbb{T}\#\mathbb{T}\#\mathbb{T}$, has Euler number

$$(\text{Euler number of } \mathbb{T}\#\mathbb{T}) + (\text{Euler number of } \mathbb{T}) - 2 = -2 + 0 - 2 = -4.$$

In fact, we can see that every time we make a connected sum with $\mathbb{T}$ the surface gains one more hole and the Euler number reduces by 2. So the torus with g holes—which can be considered as $\mathbb{T}^{\#g}$, the connected sum of g copies of the torus $\mathbb{T}$—has Euler number

$$\text{Euler number of } \mathbb{T}^{\#g} = 2 - 2g.$$

The number g of holes in the surface $\mathbb{T}^{\#g}$ is called the **genus** of the surface.

One-sided surfaces

At this point, we still can't identify the peculiar surface from Figure 14 which has an Euler number of –1. So far we've only constructed surfaces with even Euler numbers and –1 is odd. In fact, with the tori $\mathbb{T}^{\#g}$, we've only met half the story and half of the closed surfaces. Recall how in Figure 12 we made a cylinder by gluing two edges of a square. We could, instead, have glued those two sides using reversed arrows (Figure 17(a)), introducing a single twist. So the points near v_2 on e_1 are glued to the points near v_4 on e_3 and those near v_1 on e_1 are glued to the points near v_3 on e_3. This would have created a **Möbius strip**, named after August Möbius who discovered it in 1858.

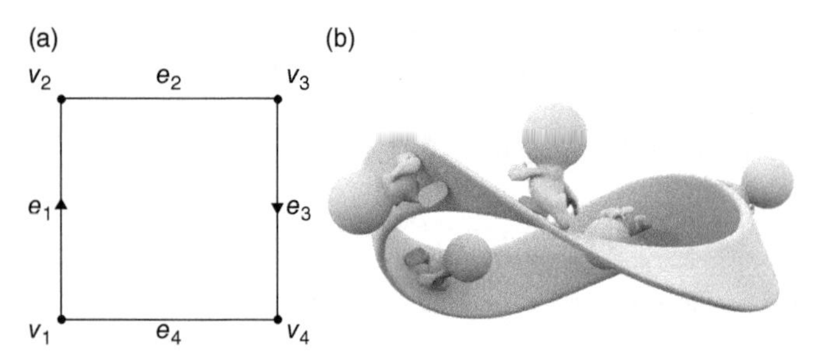

17. The Möbius strip (a) A square with identified edges, (b) Runners on a Möbius strip.

The Möbius strip is unusual in only having one side—this is apparent in Figure 17(b) as the runners cover the entirety of the strip rather than just one side of it as they would if running around just the outside (or inside) of a cylinder. Or you can imagine painting the outside of a cylinder black and the inside white, but should you begin painting a Möbius strip one colour you would find yourself covering the entire strip in that colour. The Möbius strip is an example of a *non-orientable* surface. Like the cylinder it is a surface with boundary, but note its boundary is a single circle rather than two separate ones as with the cylinder.

In Figures 18(a)–(d) we see an oriented loop—here a circle—moving around a Möbius strip. By an oriented loop I mean a loop with a given sense of direction, here initially (18(a)) appearing as clockwise to the reader. But as this loop moves around the strip (or equivalently moves left in the square) we see that when the circle returns to its original position (18(d)) that sense has now reversed and appears anti-clockwise. If you are having a little trouble visualizing what's happening to the loop, note in 18(b) and 18(c) how the points labelled P are glued together and likewise the Qs. In 18(b) most of the loop (on the left) looks to be clockwise running from P to Q, but as the loop appears on the right and continues from Q to P that sense is beginning to appear as anti-clockwise.

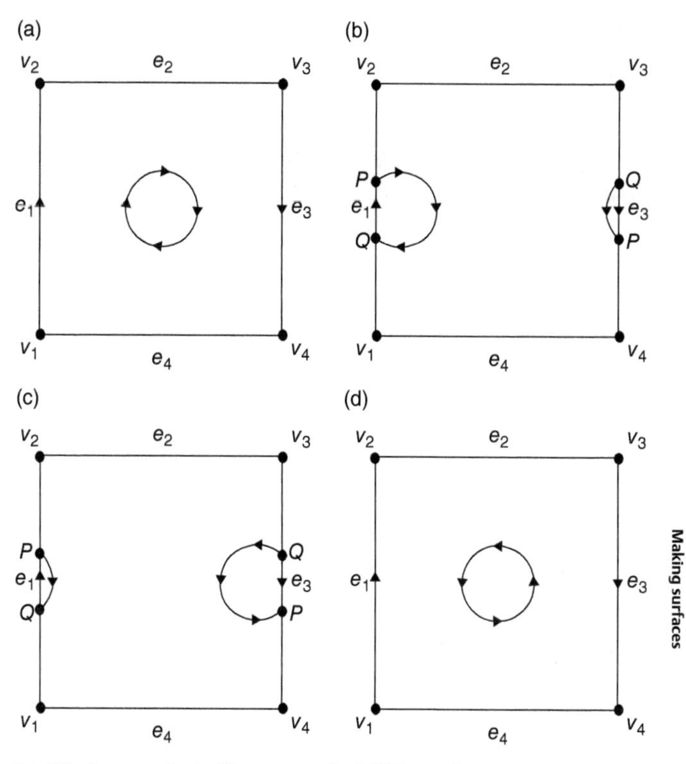

18. Moving an oriented loop around a Möbius strip.

Any surface on which it is possible to reverse the sense of an oriented loop is called **non-orientable**. If it is impossible to reverse a loop's sense, then the surface is called **orientable**. Any surface that contains a Möbius strip is non-orientable as we could just send an oriented loop once around that strip to reverse its sense. A surface with an inside and an outside is orientable. To appreciate this, imagine walking around the outside of such a surface. Looking down to your feet on the surface you could draw a circle in a clockwise manner. As you wander around the outside of the surface you can consistently take your notion of clockwise

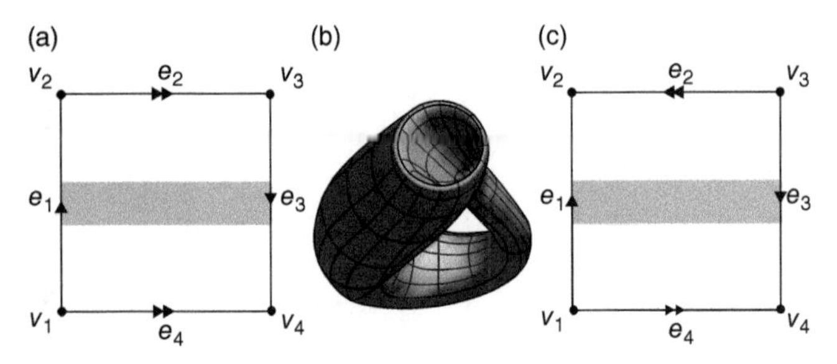

19. The Klein bottle and projective plane (a) A Klein bottle, (b) 3D depiction of a Klein Bottle, (c) A projective plane.

across the whole surface. This means, in particular, that the tori $\mathbb{T}^{\#g}$, which we met earlier and which each have an inside and outside, are all examples of orientable surfaces.

Returning to Figure 17(a), a partly glued square making a Möbius strip, there remain two unglued edges e_2 and e_4. We could glue these together as in Figure 19(a), but what surface would we make? Certainly a non-orientable one as it contains a Möbius strip (the shaded region). If instead we make this surface by gluing e_2 and e_4 first, we first create a cylinder with e_1 and e_3 as its circular ends. But to complete the surface, rather than bringing those circular ends together as with a torus, one circular end has to be glued backwards on to the other circular end—this is because of the reverse arrows on e_1 and e_3. Figure 19(b) shows how we might try to do this; we could take one circular end back into the cylinder and glue it to the other end from inside, and this way the reverse arrows line up properly. The surface made is called a **Klein bottle**, after Felix Klein who first described it in 1882. Being non-orientable, the Klein bottle does not have an inside and outside.

There is a subtle problem with the Klein bottle in Figure 19(b). When we take the cylinder back into itself, some single points in space actually represent two distinct points on the Klein bottle.

So this image is not a proper representation or *embedding* of the Klein bottle in 3D. In fact, it is *impossible* to construct a Klein bottle in 3D without such self-intersections as occur where the cylinder cuts back into itself. The relevant result demonstrating this impossibility can be viewed as a generalization of the Jordan curve theorem. That theorem concerned embedding circles in the plane with a Jordan curve having an inside and an outside. In a like manner when a closed surface is embedded in 3D, the surface again divides the remaining space into an inside and an outside and so the closed surface must be orientable. As the Klein bottle is non-orientable, it cannot be embedded in 3D.

However, the Klein bottle *can* be embedded in 4D and this isn't too hard to imagine if we treat the fourth dimension as time. The Klein bottle is two-dimensional (as surfaces are) and so from this 4D viewpoint it is important to consider the Klein bottle as only existing for an instant, a certain 'now'; for it to have a past or future would give it a third dimension. So when faced with bringing the cylinder back into itself—which would normally cause self-intersections—we can instead move that bit of cylinder gradually into the future (the fourth dimension), where the remainder of the Klein bottle doesn't exist and then, once the cylinder has passed through the space its present self occupies, we can gradually bring that bit of the cylinder back into the present. The self-intersections no longer occur, as the distinct points of the Klein bottle that became merged in Figure 19(b) instead sit in the same point of space but crucially at different times.

We can also determine the Euler number of the Klein bottle, again being careful to note how edges and vertices are glued together. The square is our only face; e_1 and e_3 are glued together, as are e_2 and e_4, making two rather than four edges; finally v_1 is glued to v_2 which is glued to v_4 which is glued to v_3 and so we have just one vertex, giving $V - E + F = 1 - 2 + 1 = 0$, the Euler number of the Klein bottle. Unfortunately, 0 is also the Euler number of the torus, so any hope we might have had that the

Euler number alone is information enough to recognize the shape of a surface was simplistic. The torus and Klein bottle are different surfaces—the former is orientable (two-sided), the latter not—and yet they both have the same Euler number.

Another important non-orientable surface, which can be formed from gluing a square's edges together, is the **projective plane** $\mathbb{P}$. In Figure 19(c) we assign e_2 and e_4 reverse arrows (in contrast to 19(a)). The surface formed is non-orientable, as it again contains a Möbius strip (the shaded region), and we can calculate the Euler number as before: again $F = 1$ and $E = 2$ but this time v_1 and v_3 are glued together and separately v_2 and v_4 are glued, so that $V = 2$. Hence $\mathbb{P}$ has Euler number $V - E + F = 2 - 2 + 1 = 1$.

The classification theorem

Classification is an important theme in mathematics. A mathematical theory often begins with definitions and rules about certain mathematical objects or structures (say functions or curves) and seeks to prove results about them using those rules. It's natural to search for examples satisfying those rules, preferably producing a complete list or *classification* of such objects.

We are now close to classifying closed surfaces. Explicitly, we are seeking to give a complete list of all the closed surfaces, so that every closed surface is homeomorphic to (i.e. topologically the same as) one of the surfaces on the list, and the list contains no duplicates—each surface on the list can be shown to be topologically different from all others on the list.

It turns out that the Euler number goes a long way to separating out the different surfaces, but we have seen that this cannot be the whole story as the torus and Klein bottle have the same Euler number whilst being different surfaces—the first is orientable, the second not. The only missing ingredient in the classification is that notion of orientability.

So the first half of the classification theorem for two-sided surfaces states:

- An orientable closed surface is homeomorphic to precisely one of the tori $\mathbb{T}^{\#g}$ where $g = 0,1,2\ldots$ These tori are not topologically the same as one another as they have different Euler numbers—the Euler number of $\mathbb{T}^{\#g}$ is $2-2g$.

A similar result holds for one-sided closed surfaces. Just as the torus $\mathbb{T}$ is a building block for the orientable surfaces, so can the projective plane $\mathbb{P}$ be used to make the non-orientable surfaces. Recall that the projective plane $\mathbb{P}$ has Euler number 1. So the connected sums $\mathbb{P}\#\mathbb{P}$ and $\mathbb{P}\#\mathbb{P}\#\mathbb{P}$ have

$$\text{Euler number of } \mathbb{P}\#\mathbb{P} = 1+1-2=0,$$
$$\text{Euler number of } \mathbb{P}\#\mathbb{P}\#\mathbb{P} = 0+1-2=-1,$$

and more generally k copies of $\mathbb{P}$ in a connected sum, a surface denoted $\mathbb{P}^{\#k}$, has Euler number $2-k$.

And the second half of the classification theorem for one-sided surfaces states:

- A non-orientable closed surface is homeomorphic to precisely one of $\mathbb{P}^{\#k}$ where $k = 1,2,3\ldots$ These surfaces are not topologically the same as they have different Euler numbers—the Euler number of $\mathbb{P}^{\#k}$ is $2-k$.

Making a connected sum with $\mathbb{P}$ is equivalent to sewing a Möbius strip into the surface. $\mathbb{P}$ itself can be made by introducing a Möbius strip into a sphere; to do this we might make a tear in the sphere and then, rather than gluing the tear back together, we could instead assign reverse arrows to the two sides of the tear, thus introducing a Möbius strip. So the surface $\mathbb{P}^{\#k}$ can be thought of as a sphere with k Möbius strips sewed in.

Overall then, the classification theorem says that if we know the Euler number of a closed surface and whether it is one- or two-sided, then we know its topological shape. If you were wondering, where the Klein bottle is on this list, we know its Euler number to be 0 and we know it to be one-sided. The only surface in the classification matching these facts is the $k = 2$ surface $\mathbb{P}\#\mathbb{P}$ and this is topologically the same as the Klein bottle. We might create a yet more complicated connected sum such as $\mathbb{T}\#\mathbb{T}\#\mathbb{P}\#\mathbb{P}\#\mathbb{P}$ which at first glance is not on our list. This surface is one-sided and its Euler number equals

$$\begin{aligned}&(\text{Euler number of } \mathbb{T}\#\mathbb{T})+(\text{Euler number of } \mathbb{P}\#\mathbb{P}\#\mathbb{P})-2\\&=(2-2\times 2)+(2-3)-2=-5,\end{aligned}$$

so topologically it's the same surface as $\mathbb{P}^{\#7}$. And at long last we are able to identify the surface we formed in Figure 14. That surface had Euler number −1 and so the surface is $\mathbb{P}\#\mathbb{P}\#\mathbb{P}$, this being the only surface on our list with that Euler number.

Complex numbers

Surfaces are a natural two-dimensional extension of one-dimensional curves which mathematicians had long been interested in but, historically, surfaces and their topology became of particular importance because of the work of 19th-century mathematicians, most notably Bernhard Riemann.

To understand Riemann's motivation for studying surfaces, we need to take a brief foray into the world of complex numbers. Complex numbers have, at first glance, nothing to do with topology, but the need to introduce them here is a consequence of the deep interconnectedness of mathematics. In the mid-19th century mathematicians found worthwhile reasons to think about older mathematics in new *topological* ways. It might then seem as though topology was somehow born of practical necessity for addressing these older problems. However I'd like to suggest a rosier picture of how mathematicians think: nothing will put a glint in the eyes of a generation of mathematicians, an itch to be thinking hard about the essence of mathematics, so much as a sense of there being something profound just around the corner and a deeper understanding of their subject tantalizingly beyond their fingertips. And so it was to prove.

These so-called 'complex' numbers arose—somewhat uncertainly—from the work of Italian mathematicians during the Renaissance. For a long time mathematicians had been interested in the solutions of *polynomial* equations. These are equations involving powers and multiples of an unknown quantity, say x, such as

$$x^3 - 2x^2 - x + 2 = 0.$$

This is a degree 3 equation, that being the highest power of x. A *solution* of an equation is a value of x which makes both sides equal. We can see that $x = 1$ solves this equation because

$$1^3 - 2 \times 1^2 - 1 + 2 = 0.$$

You might check that $x = 2$ is a solution and so is $x = -1$. And that's all of them! Three solutions $x = 1, -1, 2$. Other polynomials, though, seem to have no solutions. For example, the degree 2 equation

$$x^2 = -1$$

has no real numbers as solutions. If you take a positive number x then its square x^2 is also positive (and so cannot equal -1); if you take a negative number then its square is also positive; finally $0^2 = 0$. So there are no solutions. If you prefer a more pictorial approach then you might draw the graphs of $y = x^2$ and $y = -1$, and the fact that these graphs don't meet (Figure 20(a)) is again another way of showing that no number x solves the equation $x^2 = -1$. Basically the problem is that negative numbers don't have real square roots.

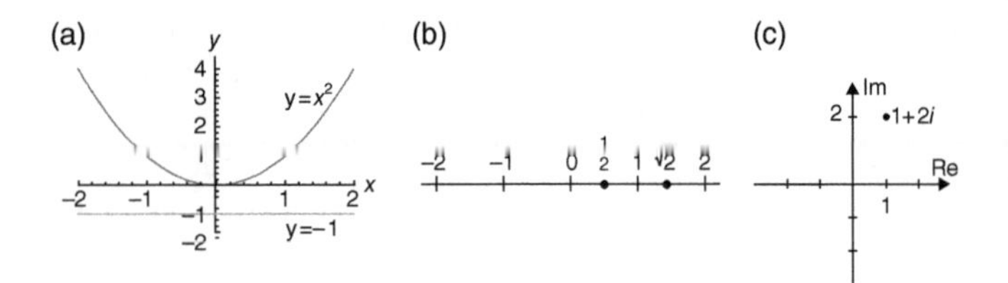

20. The real line and complex plane (a) Graphs of $y = x^2$ and $y = -1$, (b) The real line, (c) The complex plane.

And there the story might have ended except those Renaissance mathematicians found good reasons to 'imagine' that $x^2 = -1$ does have solutions, denoting a solution as i. This may seem somewhat ludicrous at first, but around 1530 a method was found for solving degree 3 equations. One problem was that this method necessitated calculations with square roots of negative numbers, even when all the equation's solutions were real numbers. The worth of the number i became truly apparent with the proof of the **fundamental theorem of algebra** in 1799 by Carl Gauss. This theorem shows *all* the solutions of *any* polynomial equation have the form $a + bi$ where a and b are real numbers. For example, the number $3 + 2i$ solves the equation

$$x^2 - 6x + 13 = 0$$

as shown by the calculation

$$\begin{aligned}&(3+2i)^2 - 6(3+2i) + 13\\ &= (9+12i+4i^2) - (18+12i) + 13\\ &= 9+12i-4-18-12i+13 \qquad [\text{recalling } i^2 = -1]\\ &= 0.\end{aligned}$$

Numbers of this form, $a + bi$ where a and b are real numbers and $i^2 = -1$, are called **complex numbers** and the fundamental theorem of algebra says that a polynomial of degree n has (counting possible repeats) n solutions amongst the complex numbers.

In the same way that real numbers are commonly represented on the *real line* (Figure 20(b)) the complex numbers can be represented as a plane, the *complex plane* (Figure 20(c)). A complex number such as $1 + 2i$ can then naturally be identified with the point (1, 2) as shown. The real numbers occupy the horizontal axis—denoted 'Re'—and the vertical axis 'Im' is called the *imaginary axis*.

Complex numbers have a rich theory of their own which, for mathematicians at least, is reason enough to warrant their study. You may, though, be surprised to find that quantum theory, the physical theory that successfully models subatomic physics, is naturally described using the language of complex numbers and so physicists, chemists, and engineers all need to be well versed in the use of complex numbers.

Riemann surfaces

The introduction of complex numbers led to a much richer theory connecting algebra and geometry. In Figure 20(a) we see that the curves $y = x^2$ and $y = -1$ don't meet; if they did meet at a point (x,y) in the real xy-plane then we'd have $x^2 = -1$ and no such x exists. But using complex numbers they do intersect at two points, at $x = i$, $y = -1$ and at $x = -i$, $y = -1$. The fact that a degree 2 curve and a degree 1 curve meet in $2 \times 1 = 2$ points in this case is not entirely coincidental. More generally it is the case that, *if properly counted*, a degree m curve and a degree n curve intersect in $m \times n$ points. Multiple contacts need to be counted properly—so a line tangentially meeting the curve $y = x^2$ would count as a double contact, so that there are still $2 \times 1 = 2$ intersections. The final finesse, when counting intersections, is to include *points at infinity*. For example, two parallel lines—each degree 1 curves—are still deemed to meet at a point at infinity so that there is $1 \times 1 = 1$ intersection as expected.

Using real numbers, the graph of $y = x^2$ is a one-dimensional curve lying in the two-dimensional xy-plane. In this case x and y are everyday real numbers and the curve consists of all points (x, x^2) where x is a real number. We might instead consider the same equation where x and y can now be complex numbers. Again all the points satisfying $y = x^2$ are of the form (x, x^2) but this time x can be any complex number. When using real numbers, the input x represents some point of the x-axis and the corresponding output x^2 can be plotted distance x^2 above the point $(x, 0)$ in the xy-plane (Figure 20(a)). However, when it comes to using complex numbers, the input $x = a + bi$ is itself two-dimensional. The x-'axis' is a version of the complex plane, the y-'axis' a second version, and the complex xy-'plane' is in fact four-dimensional. 'Above' the point $(x, 0)$ is a point (x, x^2), and together the points (x, x^2) make a two-dimensional surface situated in the four-dimensional complex xy-space. All the points such as $(2, 4)$ that were on the original real curve are still present, and make up a cross-section of the complex surface; present too now are points like $(i, -1)$ and $(2+i, 3+4i)$. If we separate out these complex numbers into their real and imaginary dimensions, then we might instead represent these points as

$$(2,4)\leftrightarrow(2,0,4,0), \quad (i,-1)\leftrightarrow(0,1,-1,0),$$
$$(2+i,3+4i)\leftrightarrow(2,1,3,4),$$

and their four-dimensional nature is a little clearer.

The curve $y = x^2$ sits in the real xy-plane as a curved version of the x-axis (Figure 20(a)); the curve and axis are topologically the same with a homeomorphism between the two just pushing each point $(x, 0)$ up to the point (x, x^2). If we include also the curve's point at infinity bringing together the curve's 'ends' then the curve topologically becomes a circle. When using complex numbers, the

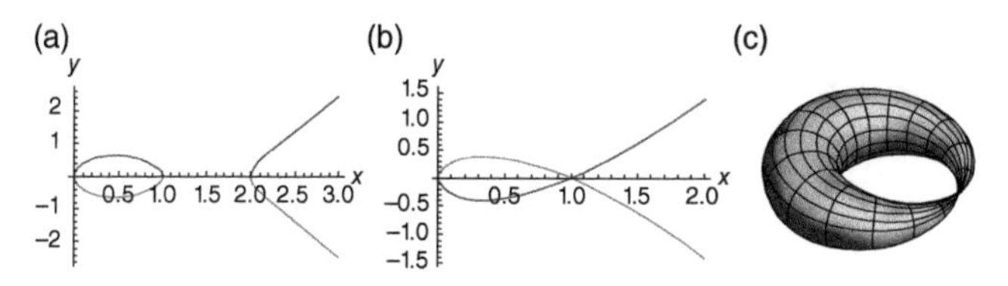

21. Visualizing Riemann surfaces (a) Graph of $y^2 = x(x - 1)(x - 2)$, (b) Graph of $y^2 = x(x - 1)^2$, (c) A pinched torus.

curve $y = x^2$ sits in complex xy-space as a curved version of the x-axis which, remember, is itself a two-dimensional complex plane. When we include the point at infinity this brings together this curved plane as a sphere. (This is the reverse process of puncturing a sphere to get the plane that we met earlier in the proof of Euler's formula.)

So, the complex version of $y = x^2$, if we include its point at infinity, is topologically a sphere, a surface; this is called the **Riemann surface** of $y = x^2$. We might similarly consider the Riemann surfaces of higher degree equations.

In Figure 21(a) we have the real cross-section described by the degree 3 equation $y^2 = x(x-1)(x-2)$. On the left is a loop, and when we add the point at infinity to the curve on the right then topologically this real cross-section becomes two loops; so it might not be surprising that the whole complex version, the Riemann surface, in this case is a torus with Figure 21(a) just being a cross-section of that torus. In Figure 21(b) we have a real cross-section with a *singular* point where the curve crosses itself. Topologically the complex version of this curve is a *pinched torus* as in Figure 21(c).

Provided there are no singular points, then a degree d equation defines a Riemann surface which is topologically a torus with g holes. There is a profound but easily described connection between the degree of a curve's equation d and the genus g of its

Riemann surface. This is given by the **degree-genus formula** which states that

$$g=\frac{1}{2}(d-1)(d-2),$$

where g is the genus of the Riemann surface and d is the degree of the curve's equation. Remembering the examples we have met, note that $d = 2$ for $y = x^2$ gives $g = 0$, a sphere, and $d = 3$ for $y^2 = x(x-1)(x-2)$ gives $g = 1$, a torus. For curves with singular points, the formula can be generalized including a correction term for each singularity, as shown by Max Noether in 1884.

This is a first glimpse at some of the deep connections between topology, algebra, geometry, and calculus that would be uncovered in mathematics. Quoting the French mathematician Jean Dieudonné, 'In the history of mathematics the twentieth century will remain as the century of topology.' The century would see a fledgling subject, intuitively but informally understood, go on to become one of the central pillars of mathematics.

It is worth noting that those early topologists—Möbius, Klein, Riemann—did not in their time have available the rigorous definitions necessary to prove their results to modern standards. In 1861 Möbius gave an early sketch proof of the classification theorem for orientable surfaces, and Walther Von Dyck gave a sketch proof for all closed surfaces in 1888. But without having any formal definition of what a surface is, these proofs can at best be considered incomplete. This is not to relegate such proofs to the dustbin, nor to consider them simply wrong, as such proofs often contain most or all of the crucial ideas of a proof. Somewhat differently expressed rigorous versions of the classification theorem would be proved by Max Dehn and Poul Heegaard in 1907 and by Roy Brahana in 1921.

Curves and surfaces are one- and two-dimensional examples of **manifolds**, spaces that look 'up close' like the real line, the plane, or some higher dimensional equivalent. It wasn't until 1936 that Hassler Whitney gave the modern definition of a manifold, and proved an important theorem showing when manifolds can be embedding in space (recall, for example, how the Klein bottle cannot be made in 3D space without self-intersections but can be made in 4D). An important aspect of modern geometry concerns the different types of mathematical structure—continuous, smooth, complex, metric—that can be put on these manifolds and I will say a little more on this in Chapter 5 when we discuss differential topology.

Chapter 3
Thinking continuously

Given just one sentence for the task, many topologists might choose to describe their subject as the study of *continuity*. The word 'continuous' appeared a few times in Chapter 1 but it was left to the reader's intuition as to quite what the word entailed. In many ways this reflects how mathematicians used to regard continuity—historically it was just considered evident what was meant by 'continuous' and as many (but not all) of the results about continuous functions that are 'obvious' also happen to be true, relatively little effort was spent providing further clarity. A rigorous definition of continuity did not appear until the 19th century.

In your everyday routine there are continuous and discontinuous functions around you. For example, if you drive to work, the distance you have travelled after a certain time will be a continuous function of time—for this not to be the case would mean that at one moment your car was in a certain place only for it to immediately afterwards be at another place some distance away. Your speed on the journey will similarly be continuous. However, the acceleration need not be; if you were sat at rest (say at traffic lights) the acceleration would be zero but then would jump to a certain value once your foot was on the accelerator. The graphs in Figure 22 give a plausible (if simplistic) model for someone's drive to work.

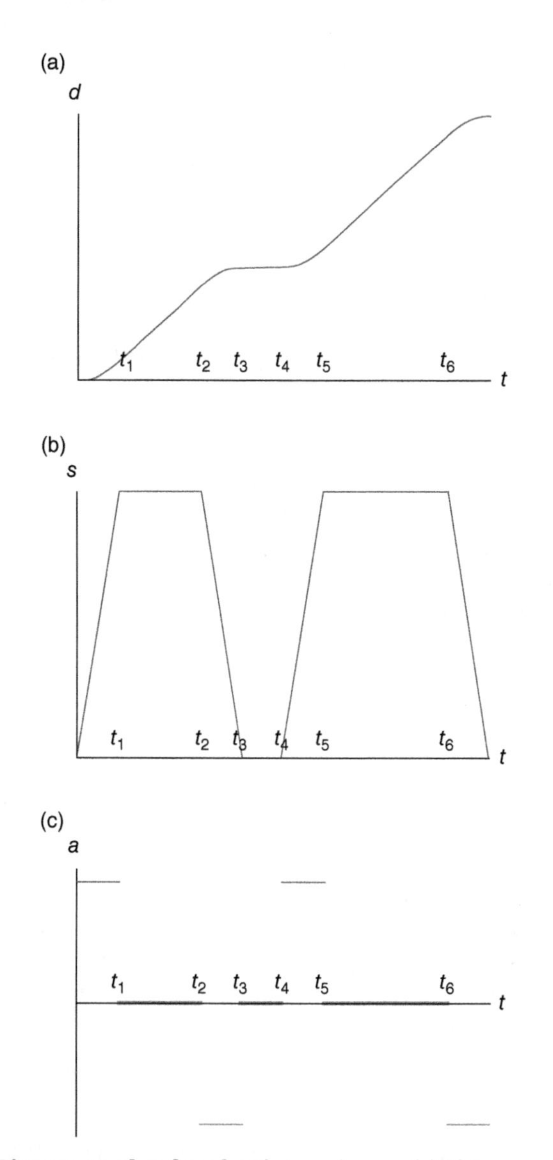

22. Distance, speed, and acceleration on a journey (a) Distance, (b) Speed, (c) Acceleration.

From Figure 22(b) we can see that the car stops at t_3—where the speed $s(t)$ becomes zero—and after t_4 increases to the speed limit. The distance travelled $d(t)$ in Figure 22(a) is a continuous function of time t. Historically this would have been understood as meaning its graph could be drawn without taking pen from paper, but we will seek to provide a fuller understanding. But the acceleration function $a(t)$ is not continuous because of the jumps in the graph in Figure 22(c). The times $t_1, t_2, \ldots t_6$ of discontinuity in the acceleration relate to the driver's foot coming off the accelerator, being put on the brake, coming off the brake, and then the pattern repeats again.

My aim in this chapter is to provide a more rigorous sense of just what continuity entails for *real-valued functions of a real variable.* This means we will focus on functions having a single numerical input and a single numerical output.

Functions

The idea of a function is a central one to mathematics, though this has only been true since around the 17th century. Once Descartes and Fermat independently introduced the idea of Cartesian coordinates x and y to describe position in a plane, a curve could just as easily be described by an equation as by its geometry. For example, the curve $y = x^2$ is a parabola, a curve the ancient Greeks would have investigated solely using geometry. A sketch of this curve is given in Figure 23(a)—the curve's equation gives a rule for plotting, above each point $(x, 0)$ of the x-axis, a point (x, x^2). Note how certain algebraic properties of the function are represented in the shape and position of the curve—as $x^2 \geqslant 0$ for all x, the curve lies entirely on or above the x-axis; as $(-x)^2 = x^2$, the curve is symmetric about the y-axis.

For some functions, we might naturally have to limit the allowed inputs—or we might choose to do so anyway. For example, if $f(x) = 1/x$ then we at least need to ensure that x is non-zero as division by zero is meaningless; for the function $f(x) = \sqrt{x}$, then

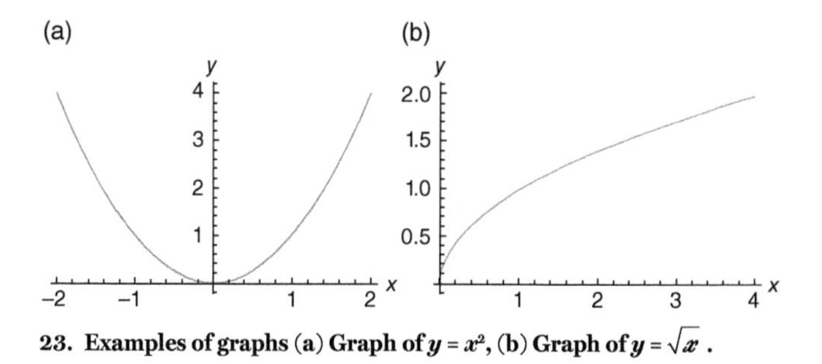

23. Examples of graphs (a) Graph of $y = x^2$, (b) Graph of $y = \sqrt{x}$.

we cannot permit x to be negative, as no real number has a negative square (Figure 23(b)).

More generally, a function comes with a set of inputs, known as the *domain*, and there is likewise the *codomain*, a set containing the outputs. It is an important, if subtle, point to appreciate that a function is this whole package: the domain, the codomain, and the rule assigning values.

Some first thoughts about continuity

Let's first try to understand what it means for a function, with real inputs and outputs, to be continuous. Currently we sort of intuitively know continuity when we see it. Certainly, looking at two functions in Figures 24(a) and 24(b), it seems reasonable to say $f(x)$ is continuous and $g(x)$ is not continuous, and further that $g(x)$ is discontinuous only at $x = 1$. (The full disc on the graph shows where the function takes its value, so that $g(1) = 1$.) But what does intuition say about Figure 24(c)? Is $h(x)$ continuous or not? It seems that, if $h(x)$ is discontinuous, the only point of discontinuity is $x = 0$, but the function oscillates so wildly there, we may now be thinking that our intuition didn't have all the answers.

Back to Figure 24(b), what is it about the function's behaviour at $x = 1$ that makes us think $g(x)$ is discontinuous? For input x a little

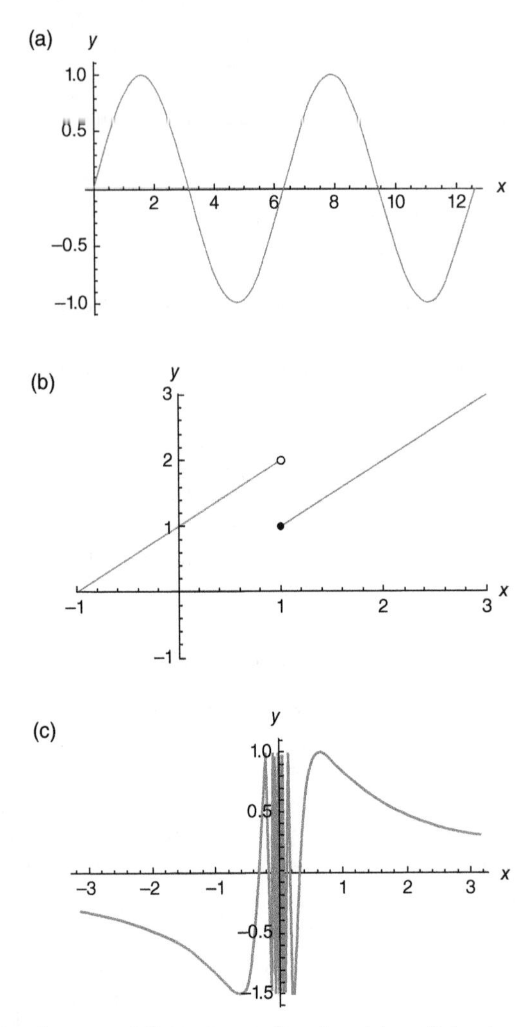

24. Continuous and discontinuous functions (a) $y = f(x) = \sin x$, **(b)** $y = g(x)\begin{Bmatrix} x+1 \text{ for } x<1 \\ x \text{ for } x \geqslant 1 \end{Bmatrix}$, **(c)** $y = h(x) = \begin{cases} \sin(1/x) & \text{for } x \neq 0 \\ 0 & \text{for } x = 0 \end{cases}$.

more than 1, then $g(x)$ has much the same value as $g(1)$; however for input x a little less than 1 then $g(x)$ is noticeably different from $g(1)$. It is this jump in output at 1 that is crucial to $g(x)$ being discontinuous at $x = 1$.

At first, we might be tempted to think this is because $g(1)$ is different from the value of $g(x)$ achieved *immediately before* we get to x equalling 1. But there are all sorts of problems with this thinking. First, there is no real number x that is 'immediately before' 1. Given a number like 0.999, close to 1, then we can always improve on that and see 0.9999 is a little closer. Or we might suggest using 0.999... (where the ellipsis means that there are infinitely many recurring 9s) but this is just another decimal expansion for 1. More rigorously, for any input $x < 1$ then $(1+x)/2$ is less than 1 but closer to 1 than x is. Instead we might be tempted to talk about an input that is infinitesimally close to 1 but then—whatever we mean by this—we are no longer talking about the real numbers and have just replaced resolving one definition with resolving a different one.

We need another approach that can be comfortably expressed entirely in terms of real numbers. This problem was independently resolved in the 19th century by Bernard Bolzano and Karl Weierstrass. We feel that $g(x)$ is discontinuous at $x = 1$ because $g(x)$ is noticeably different from $g(1)$ for some inputs x nearby to 1.

There is still quite a bit of subtlety needed to fully capture what this means. In our example, $g(x)$ has a jump of 1 from output values near 2 (just before $x = 1$) to output values near 1 (just after $x = 1$). The size of that jump was unimportant, the presence of any jump at all was sufficient. And the notion of 'nearby inputs' should not be interpreted as several inputs that are in some sense close to 1; rather we mean there are inputs x arbitrarily close to 1 such that $g(x)$ is noticeably different from $g(1)$. Necessarily this means that we are talking about infinitely many such inputs x, not just several x. By way of example, it is enough to note that:

$$
\begin{aligned}
&g(0.9) = 1.9; \\
&g(0.99) = 1.99; \\
&g(0.999) = 1.999; \\
&g(0.9999) = 1.9999; \\
&\vdots \\
&g(1) = 1.
\end{aligned}
$$

This rigorously shows that $g(x)$ is discontinuous at $x = 1$. The sequence of inputs 0.9, 0.99, 0.999, 0.9999, ... gets arbitrarily close to 1. What this means is: however demanding 'nearby to 1' is required to be, there are inputs from this sequence that are at least that close.

Whilst we still haven't quite defined just what we mean by discontinuous, we have made some progress with regard to the function $h(x)$ (Figure 24(c)). This function does not appear to have any noticeable 'jump' in outputs, but it does seem to meet the definition

> $h(x)$ is noticeably different from $h(0)$ for some inputs x arbitrarily near to 0.

Near $x = 0$ the function $h(x)$ is varying crazily. From the graph we can see that there are inputs x, arbitrarily close to 0, where $h(x) = 1$ whilst we have $h(0) = 0$. It now seems clear by our emerging sense of continuity that $h(x)$ is discontinuous at $x = 0$.

An example in detail

We still need to be careful turning these nascent thoughts into a rigorous definition. We'll consider in detail the function $f(x) = x^2$

which *is* continuous for all inputs x. If $f(x)$ is continuous at an input $x = a$ then—based on our previous thoughts—we need that

> $f(x)$ is not noticeably different from $f(a)$ for all inputs x suitably near to a.

Take a moment to appreciate why we need all inputs x suitably near to a to produce not noticeably different outputs $f(x)$ to $f(a)$. If some—but only one—nearby input x to a resulted in noticeably different outputs $f(x)$ and $f(a)$, then we could just tighten our notion of 'suitably near' to exclude the problem input x. In fact, if we can never get 'suitably near' with our inputs, then this means that there were arbitrarily close problematic inputs x to a where $f(x)$ was noticeably different from $f(a)$—so, $f(x)$ would be discontinuous at $x = a$.

To begin, what does it mean for $f(x) = x^2$ to be continuous at $x = 0$? Is it true that

> x^2 is not noticeably different from $0^2 = 0$ for all inputs x suitably near to 0?

We try out some values in Table 3.

It seems—admittedly only on the basis of five choices of x—that x^2 is closer to 0 than x is to 0, and some quick algebra checks that

Table 3. Sample input and output values for $f(x) = x^2$

x	difference from 0	x^2	difference from 0^2
0.1	0.1	0.01	0.01
0.01	0.01	0.0001	0.0001
0	0	0	0
−0.01	0.01	0.0001	0.0001
−0.1	0.1	0.01	0.01

small numbers generally square to smaller numbers (in magnitude). We cannot find inputs x close to 0 where the outputs x^2 and 0 are noticeably different.

Now we can hang some rigorous mathematics on these initial thoughts: whatever potential 'noticeable difference' in the outputs x^2 and $0^2 = 0$ we consider, represented by a positive number e, then there need to be 'suitably close' inputs x to 0, represented by a positive number d, such that

> if inputs x and 0 differ by less than d then outputs x^2 and 0 differ by less than e.

As the outputs here are closer to one another than the inputs are—that is, as x^2 is closer to 0 than x is to 0—then we can just choose d to equal e. So if inputs differ by e or less so do the outputs. We have then shown that $f(x) = x^2$ is continuous at $x = 0$.

What about continuity at a different input, say $x = 1000$? We can create a similar table to Table 3 (see Table 4).

The function $f(x) = x^2$ is growing much more rapidly at $x = 1000$ than it is at $x = 0$. A change of around 0.1 in the inputs leads to a change in the outputs of around 200; a change of 0.01 in the inputs still leads to a difference of around 20 in the outputs.

Table 4. More sample input and output values for $f(x) = x^2$

x	difference from 1000	x^2	difference from 1000^2
1000.1	0.1	1000200.01	200.01
1000.01	0.01	1000020.0001	20.0001
1000	0	1000000	0
999.999	0.01	999980.0001	19.9999
999.99	0.1	999980.01	199.99

This may lead you to think that the outputs are 'noticeably different' here, but a more careful check of other inputs would show that these large differences have been incrementally achieved. All this is a consequence of the function changing more rapidly near $x = 1000$, and what needs tightening is our notion of 'suitably near'. As the function is growing more rapidly, small changes in the input will lead to relatively large changes, but still in a continuous fashion. If we consider the input $x = 1000 + d$, a little larger than the input 1000, then the difference in the outputs equals

$$\begin{aligned}(1000 + d)^2 - 1000^2 &= \left(1000^2 + 2000d + d^2\right) - 1000^2 \\ &= 2000d + d^2 < 2001d,\end{aligned}$$

as $d^2 < d$ when $d < 1$. So a shift in inputs by d results in a shift of outputs roughly 2000 times larger. (Note similar behaviour in Table 4.) This is, in itself, not a problem but it does mean that if we want the outputs to differ by no more than e then we should only allow the inputs to differ by no more that $e/2001$. This still shows the continuity of $f(x) = x^2$ at $x = 1000$, we just needed a tighter sense of 'suitably near' with the inputs as the function was growing so fast. For continuity at yet larger inputs that notion would have to become yet more stringent, but we would always be able to find some small wiggle room about an input for which the outputs don't differ beyond the desired amount e.

A rigorous definition

Putting all this thinking together gives us a rigorous definition of continuity. I'd suggest reading the definition and seeking to understand how this means that the function in Figure 25(a) *is* continuous and the one in Figure 25(b) *isn't*, but if you find the generality of the definition and the technical level of the language difficult then move on to the next section on the properties of continuous functions. And be reassured, as it took generations of

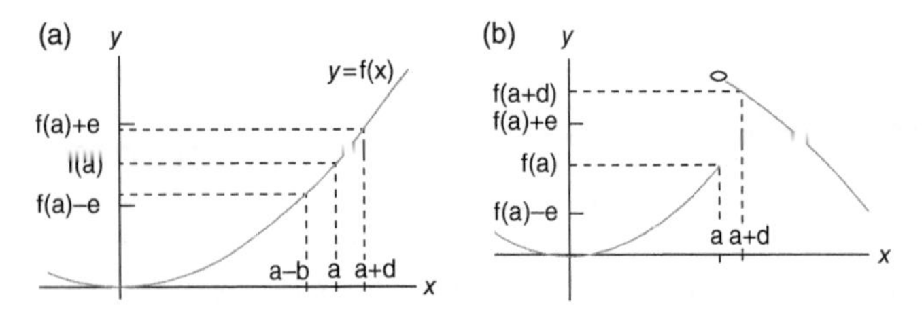

25. The rigorous definition of continuous and discontinuous (a) A continuous function, (b) A discontinuous function.

mathematicians to finally get this definition right, and current and past generations of mathematics undergraduates still wrestle with proofs involving this definition in their analysis courses.

Formally, then, a function with real inputs x and real outputs $f(x)$ is **continuous** at an input $x = a$ if:

> for any positive e there is some positive d
>
> such that the difference between the outputs $f(x)$ and $f(a)$ is less than e
>
> when the difference between the inputs x and a is less than d.

In Figure 25(a), we are focusing on demonstrating the continuity of $f(x)$ at input $x = a$. A particular choice of $e > 0$ has been made and our task now is to make sure that the outputs don't differ from $f(a)$ by more than this e. So the outputs have to remain below $f(a) + e$ and above $f(a) - e$ (as shown on the y-axis). And this has to happen for inputs x in some range $a - d < x < a+d$. We can see from Figure 25(a) that some such interval has been found, as shown on the x-axis—the range of outputs on this interval are bounded by the dashed lines and these fall within the permitted range for the outputs. To show continuity of $f(x)$ at an input $x = a$ we'd have to show that this can be done for *all* $e > 0$, however small; to show continuity of the function $f(x)$ we'd have to do this for *all* inputs x.

There are several important points to note here:

- we require that the outputs can be constrained in a certain way if the inputs are appropriately constrained;
- we need to be able to do this for *all* constraints e in the outputs; for each choice of e we will need a choice of d that meets the requirement;
- for a smaller choice of e then d will usually need to be smaller as well;
- given a positive e, then any positive d that meets the requirement is fine—we're not looking for a largest such d, say;
- the faster the function $f(x)$ is changing at a, the smaller d will need to be relative to e.

A function is then said to be continuous if it is continuous at all its inputs. And for a function $f(x)$ to be **discontinuous** at an input $x = a$ means:

> there is some positive e such that for any positive d
>
> the difference between the outputs $f(x)$ and $f(a)$ is greater than e
>
> for some input x where the difference between x and a is less than d.

Note how this captures there being inputs x arbitrarily close to a (as they can be found within any distance d of a) for which the inputs $f(x)$ and $f(a)$ are 'noticeably different' (here meaning differing by more than e). In Figure 25(b) if we choose e to be smaller than the jump in the output that occurs at $x = a$ then outputs $f(a + d)$ will be greater than $f(a) + e$ and so outside the permitted range—no matter how small we make d.

Properties of continuous functions

There would be limited reason to be interested in continuous functions if there wasn't some payback in what can be guaranteed about continuous functions, when compared with what can be said of functions in general. The **intermediate value theorem**, as you might guess from the name, states: if we have a continuous function $f(x)$ that is negative at some input $x = a$ and positive at some later input $x = b$, there is some input $x = c$ between a and b—possibly more than one—where $f(c) = 0$. Zero here is the eponymous 'intermediate value' between the negative starting output $f(a)$ and the final positive output $f(b)$.

If $f(x)$ is continuous, then this seems like something that just *has* to be true: if we were to draw the graph of $y = f(x)$ between the point $(a, f(a))$ below the x-axis and $(b, f(b))$ above the x-axis, without taking pen off paper, we surely must cross the x-axis at least once. But, now that we have a formal definition of continuity, is it particularly clear how we would go about *proving* this result? It's not hard to come up with a counter-example to the result when the function is discontinuous. Such a function (with $a = 0$, $b = 2$, say) is

$$f(x) = \begin{cases} -1 & \text{for } 0 \leqslant x \leqslant 1 \\ 1 & \text{for } 1 < x \leqslant 2 \end{cases}$$

which is sketched in Figure 26(b). Note that its only outputs are −1 and 1, so that there are definitely no solutions to $f(x) = 0$. But to prove the intermediate value theorem, we would need to show that a function $f(x)$ satisfying the theorem's requirements takes the value zero somewhere—importantly we would know very little specifically about $f(x)$ save that it is continuous, begins negatively, and finishes positively, so any approach to a proof would have to be similarly general. Visualizing drawing the graph from beneath to above the x-axis, it seems as though there would have to be a first time that we cross the x-axis and this is indeed the case. That input value c where we first cross would be

$$c = \text{minimum value of } x \text{ such that } f(x) \geqslant 0.$$

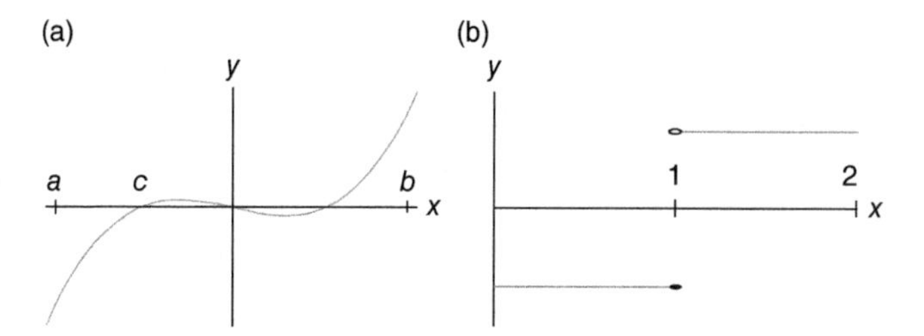

26. The intermediate value theorem (a) Intermediate value theorem, (b) A discontinuous function.

But writing down this definition of c doesn't itself constitute a proof and carefully proving the intermediate value theorem is well beyond the aims of this book. The above line is the proof's starting point, we have a candidate input c where we think the function is zero, but it remains to carefully show that $f(c) = 0$. The intermediate value theorem was first proved by Bolzano in 1817.

Another important theorem is the **boundedness theorem**. A function $f(x)$ is said to be **bounded** if there are bounds M and N such that $M \leqslant f(x) \leqslant N$ for all x. Continuous functions—like $f(x) = x^2$—can be unbounded (here no such N exists) when we consider *all* possible inputs x, but when we restrict the inputs to a domain such as the interval $a \leqslant x \leqslant b$ then the outputs $f(x)$ will be bounded. More than that, there will be a maximum output and a minimum output which are achieved at some inputs. For example, the function $f(x) = x^2$ on the interval $-1 \leqslant x \leqslant 2$ has a maximum output of 4 achieved at $x = 2$ and a minimum output of 0 achieved at $x = 0$. On the other hand a function like

$$g(x) = \begin{cases} 0 & \text{for } 0 = x \\ 1/x & \text{for } 0 < x \leqslant 2 \end{cases}$$

is not bounded on the interval $0 \leqslant x \leqslant 2$. As x becomes small, but remains non-zero, then $g(x) = 1/x$ becomes arbitrarily large. The crucial point here is that $g(x)$ is not continuous, specifically at

$x = 0$, and so the boundedness theorem does not apply in this case. The boundedness theorem was first proved by Weierstrass during the 1840s, though the result did not become widely known until he began lecturing in Berlin in 1859.

The continuous functions form an important part of mathematics because of powerful results, like the intermediate value theorem and boundedness theorem, guaranteeing certain properties for continuous functions. There are also theorems that guarantee the continuity of functions that can be constructed from other functions: if $f(x)$ and $g(x)$ are continuous functions then so are

$$f(x)+g(x),\ f(x)-g(x),\ f(x)\times g(x),\\ f(x)/g(x)\ (\text{provided } g(x)\neq 0), f(g(x)).$$

Consequently, the continuous functions include many of the functions that mathematicians and scientists routinely meet and work with. For example, any function with a defined gradient will also be continuous. However, be aware that there are still some nasty, pathological examples amongst the continuous functions—for example, the blancmange function (Figure 27(d)) is a continuous function that has a defined gradient at *no* input at all.

The blancmange function can be defined by adding together an infinite list of functions $f_1(x), f_2(x), f_3(x), \ldots$ The first three functions appear in Figures 27(a), 27(b), 27(c). Note that $f_1(x)$ has a defined gradient/slope at each point except its two peaks and the trough in the middle. There are more inputs—but still finitely many—where the second and third functions don't have a defined gradient. Rather astonishing though if we add the *whole* list of functions $f_1(x), f_2(x), f_3(x), \ldots$ then we arrive at the blancmange-shaped function drawn in Figure 27(d) and this hasn't a defined gradient at *any* point of its graph.

To conclude, recall the earlier comment that a function is the whole package of domain, codomain, and the assignment rule. We cannot

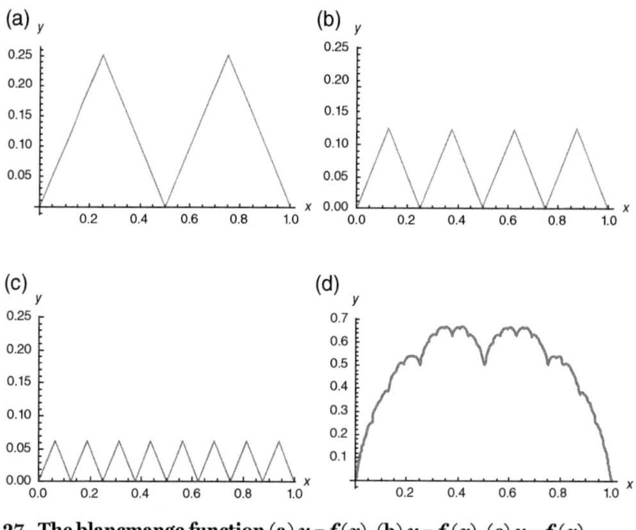

27. The blancmange function (a) $y = f_1(x)$, (b) $y = f_2(x)$, (c) $y = f_3(x)$, (d) Blancmange function.

simply say whether x^2 is a bounded function. The assignment x^2 is part of an unbounded function when the domain and codomain are both the real line, but when we restrict the domain to the interval $a \leqslant x \leqslant b$ then the assignment x^2 yields a bounded function. This pre-empts somewhat a more detailed discussion for Chapter 4: what is it about the domain $a \leqslant x \leqslant b$ that means continuous functions are bounded on that domain or satisfy the intermediate value theorem? The answer to the first question is that the interval is *compact* and to the second question is that the interval is *connected.* We will see in Chapter 4 what these terms mean more generally.

Chapter 4
The plane and other spaces

More on functions

In Chapter 3 we discussed the continuity of functions that take one numerical input and produce one numerical output. Most functions are not of such a simple form. As you read this, the density of matter in the room around you is a function of three spatial coordinates (needed to describe a point of the room you're in) and one coordinate describing time. If you are reading the paperback version, that density function takes a roughly constant value (the density of paper) at points in the book you're reading but that value changes (discontinuously) at the book's edges and takes on a new value (the density of air) at points outside the book. If you're reading this outside, the wind's velocity is an output with three components measuring to what extent the wind is blowing ahead/behind, up/down, left/right; each of these three components is again a function of one temporal and three spatial inputs. We'd expect wind velocity to be a continuous function, even if it may sometimes change quite quickly. To have you thinking a little harder, is it reasonable to say that the distance a car has travelled is a continuous function of its speed? This really is a subtle question as neither the input nor output are numbers, but rather functions of time, with input the speed function $s(t)$ and output the distance function $d(t)$ (Figure 22). If you have some sense of what the question is asking—will my journey to work tomorrow be much the same as my journey today if I keep to much the same speed during my journey?—then your intuition is doing well. Even then, there are important details to be filled in, namely describing in each case just what 'much the same' means.

Thoughts on distance

Consider the important notions needed to define continuity in Chapter 3. Loosely put, continuity requires that we can constrain the difference in outputs by suitably constraining the difference in inputs. We will need a more general notion of the difference between other types of input and output: for example, what is the 'difference' between two points of the plane?

Given two real numbers, x and y, we denote the difference between them as $|x-y|$, and this is just another expression for the distance between x and y as points on the real line (Figure 20(b)). Likewise, given two points (x_1, y_1) and (x_2, y_2) in the plane, we might take the 'difference' between them to mean the straight-line distance between them (Figure 28(a)). Pythagoras' Theorem tells us this distance equals

$$\text{straight-line distance} = \sqrt{(x_1 - x_2)^2 + (y_1 - y_2)^2}$$

so that the points (1,1) and (3,2) in Figure 28(a) are a distance $\sqrt{2^2+1^2} = \sqrt{5}$ apart. But in some circumstances, you might decide that the straight-line distance is not the best way to describe distance realistically. For example, a taxi driver in Manhattan would be constrained by New York's grid of streets and so would need to take a journey as in Figure 28(b), travelling along perpendicular streets and avenues. That distance is given by the formula

$$\text{taxicab distance} = |x_1 - x_2| + |y_1 - y_2|.$$

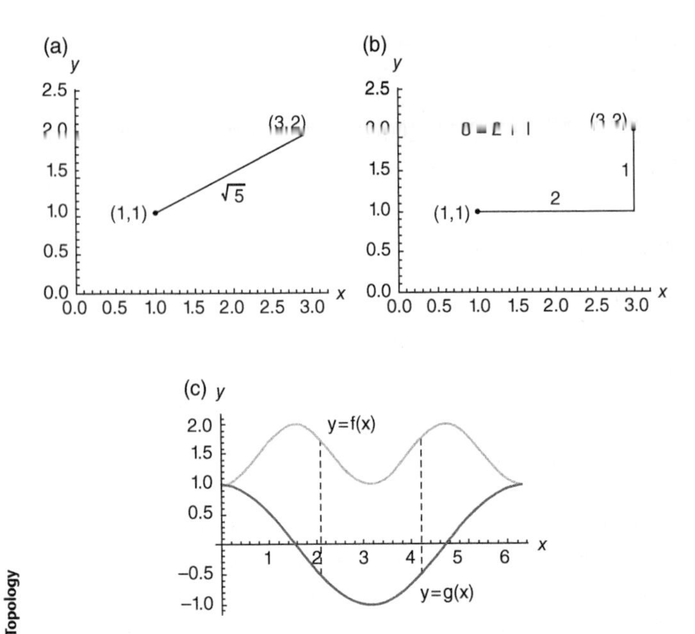

28. Visualizing different metrics (a) Straight line distance, (b) Taxicab distance, (c) Distance between functions.

For the taxi driver, the points (1,1) and (3,2) are a distance $2 + 1 = 3$ apart, as the taxi may not go along the straight path through Manhattan's skyscrapers.

A more complicated example appears in Figure 28(c): what might we mean by the distance between two functions? Again, we might choose different ways to measure that distance; looking at the two functions $f(x)$ and $g(x)$ graphed we might define the distance between them to be

> the maximum distance between $f(x)$ and $g(x)$ as we range over all inputs x,

as highlighted by the two dotted lines. But another reasonable notion of the distance between the functions might be the total area between the graphs of $f(x)$ and $g(x)$.

Again, what makes the better definition of distance between two functions may depend on what the functions signify.

Whilst these examples might seem reasonable definitions for distance, we are going to need a more detailed, firmer sense of the properties of distance before any mathematical theory can be produced. What do we expect of distance? In answering this, we find ourselves defining what it is to be a *metric*.

Sets and metric spaces

Metric spaces were first introduced by Maurice Fréchet in 1906, though his work was little appreciated at the time. A metric space is a set M together with a distance function d called the metric. This function d has two points x and y from M as its inputs and its outputs a number $d(x, y)$ representing the distance from x to y. Further the metric d must satisfy the following properties for all x, y, and z in M

1. $d(x, y) \geqslant 0$;
2. $d(x, x) = 0$ and if $d(x, y) = 0$ then $x = y$;
3. $d(x, y) = d(y, x)$;
4. $d(x, z) \leqslant d(x, y) + d(y, z)$.

The first property says that distances can't be negative and the second implies that the distance between two distinct points is positive. The third says that the distance going from one point to another is the same as the distance coming back.

The fourth property is called the *triangle inequality*. In a triangle, the combined length of two sides is always greater than the length of the third side. Similarly property 4 says that the distance $d(x,z)$ between two points x and z is never more than the distance from x to z via a third point y, namely $d(x, y) + d(y, z)$. These combined distances might be the same if y is in some sense 'on the way' from x to z but typically the diversion to y will make for a journey of greater combined distance.

Figure 28's examples are all metrics—straight-line distance and taxicab distance are metrics on the plane; maximum-distance-apart and area-between both define metrics for continuous functions on an interval $a \leqslant x \leqslant b$. Other metrics are common in mathematics, an important metric in coding and information theory being **Hamming distance**, introduced by Richard Hamming in 1950. Hamming distance is a metric for binary strings of 0s and 1s, called *codewords*, used in coding and is useful when dealing with errors when transmitting code. Given three codewords of the same length (here length 8),

$$w_1 = 11101100, \quad w_2 = 01101101, \quad w_3 = 00001000,$$

should we consider these codewords as being close to one another? Note w_1 and w_2 differ in only the first and last digits, whilst w_1 and w_3 differ in four positions as do w_2 and w_3. It seems reasonable to say w_1 and w_2 are closer, as being less distinct, and the Hamming distance between two codewords is defined to be the number of positions where the two codewords differ, so that w_1 and w_2 are distance 2 apart.

If codewords are communicated across distance, there may be a chance of transmission error. We might hope that the chance of two errors occurring while sending a single codeword is small enough to be considered negligible, but single errors might occur. It may be that the string 11101101 is received, which agrees with none of the codewords and is frustratingly Hamming distance 1 from both w_1 and w_2. The receiver would not know whether w_1 had been transmitted and the last digit mis-sent or w_2 had been sent with the first digit wrongly received. However, if we make sure to use codewords that are distance 3 or more apart, and no more than a single error takes place per codeword, then the receiver can correct the error by replacing the mis-sent code with the nearest codeword.

Continuity between metric spaces

Continuity for a function with a single numerical input and output requires that we can constrain the difference in outputs to any degree by constraining the difference in the inputs. And recall that the difference $|x-y|$ between two numbers x, y is also the distance between them on the real line. To generalize the definition of continuity to functions between metric spaces, we need to replace that previous notion of difference with that of distance, as given by some appropriate metric.

More specifically, if we have a function $f(x)$ taking inputs in a metric space M with outputs in a second metric space N, then $f(x)$ is continuous at input $x = a$ if

> for any positive e there is some positive d
>
> such that the distance in N between the outputs $f(x)$ and $f(a)$ is less than e
>
> whenever the distance in M between the inputs x and a is less than d.

Note here that the input x need no longer be a single numerical input: x could be a point in the plane, a subset of the plane, or a function itself, amongst other possibilities, and likewise there is no restriction on the output $f(x)$ being a single number.

As we have seen there are different choices of metrics on the same set—for example straight-line or taxicab on the plane—and two different metrics might have quite different views of whether two points are close to one another or not. So it is quite possible that the *same* function between two sets might be continuous when using certain metrics, and discontinuous for a *different* choice of metrics.

Before generalizing overly, let's aim to understand what it means for a function $f(x, y)$ to be continuous when there is an input (x, y) from the plane and the output $f(x, y)$ is a single real number. We'll use straight-line distance in the plane (Figure 28(a)) for the inputs and the usual notion of distance (or difference) for the outputs.

Here are three such functions.

$$f(x,y) = x^2 \qquad g(x,y) = \begin{cases} -1 & \text{for } x \leqslant 0 \\ 1 & \text{for } 0 < x \end{cases}$$

$$h(x,y) = \begin{cases} x^2 + y^2 & \text{for } x^2 + y^2 \leqslant 1 \\ 1 & \text{for } x^2 + y^2 > 1 \end{cases}$$

with their graphs sketched in Figures 29(a)–(c).

Based on the graphs, you may think that $f(x, y)$ and $h(x, y)$ are continuous, whilst $g(x, y)$ is discontinuous—all of which is true. You may further suspect that $g(x, y)$ is discontinuous at the points $(0, y)$ on the line $x = 0$, which is again correct. The jump in output $g(x, y)$ across the line $x = 0$, as we move between the two rules that define the function, is precisely why $g(x, y)$ is discontinuous there. $h(x, y)$ is also defined by two separate rules, one for inputs where $x^2 + y^2 \leqslant 1$, one for when $x^2 + y^2 > 1$. The graph of $h(x, y)$, when $x^2 + y^2 \leqslant 1$ is the bowl-shaped part of Figure 29(c). As we move towards the boundary of that rule's application, the circle $x^2 + y^2 = 1$, then $h(x, y) = x^2 + y^2$ gets ever closer to equalling 1 which is the rule for $h(x, y)$ outside of the disc. So whilst $h(x, y)$ is defined by two rules, the rule for $x^2 + y^2 \leqslant 1$ nicely hands over to the second rule for $x^2 + y^2 > 1$ without any jump in the output. This is why $h(x, y)$ is continuous.

As before there are nice algebraic results guaranteeing that if $f(x, y)$ and $g(x, y)$ are continuous functions from the plane to the real numbers then so are the functions

$$f(x,y)+g(x,y), \quad f(x,y)-g(x,y), \quad f(x,y)\times g(x,y),$$
$$f(x,y)/g(x,y) \ (\text{provided } g(x,y) \neq 0).$$

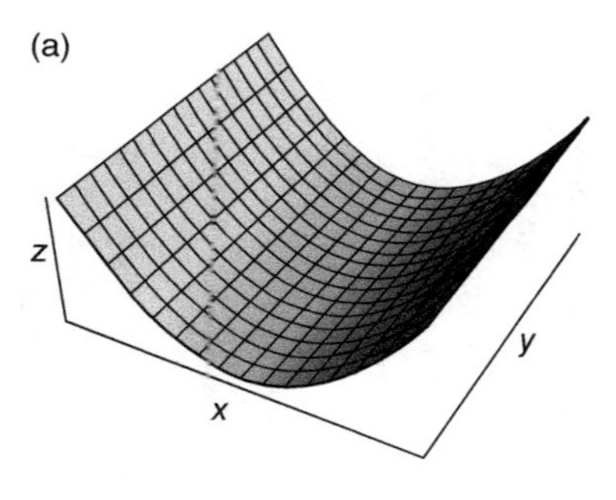

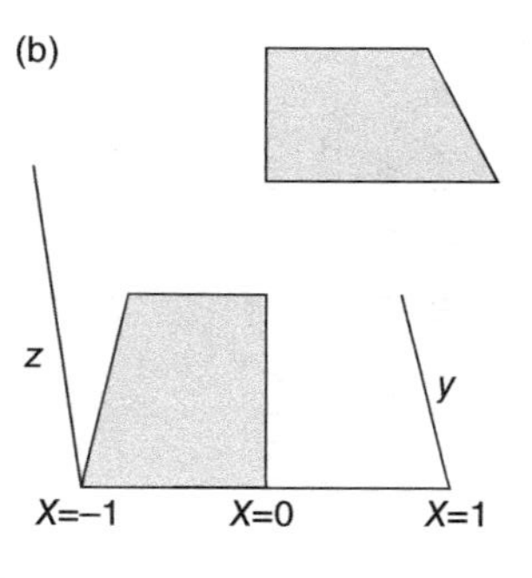

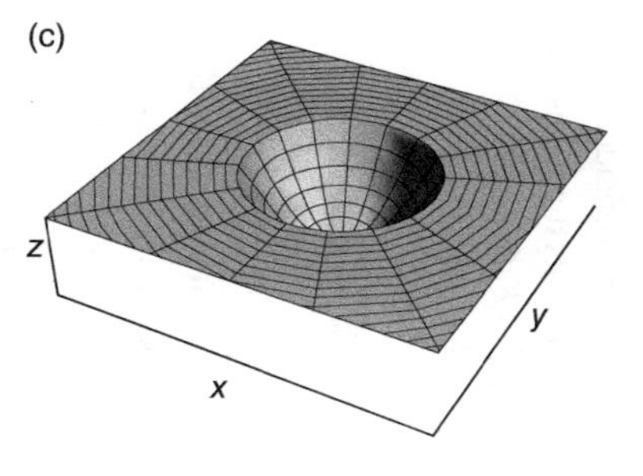

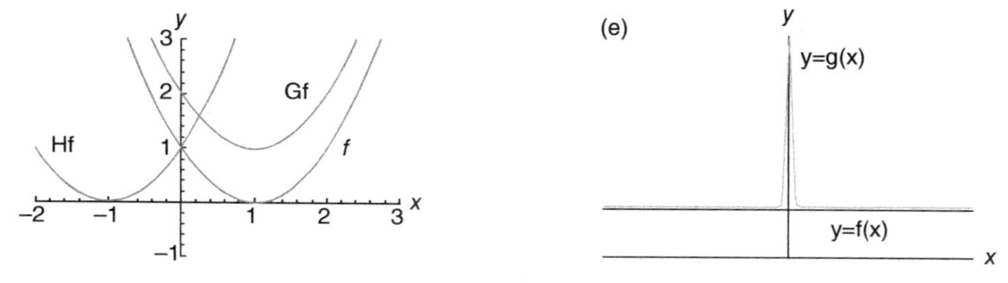

29. Graphs of functions of two variables (a) Graph of $z = f(x,y)$, (b) Graph of $z = g(x,y)$, (c) Graph of $z = h(x,y)$, (d) Eects of G and H, (e) Close and distant functions.

With the general approach of metric spaces, we can consider more complicated examples, such as the set of continuous functions on the interval $-1 \leqslant x \leqslant 1$. Here are three examples F, G, H of functions which take inputs $f(x)$ that *themselves are functions.*

$$F(f(x)) = f(0), \; G(f(x)) = f(x)+1, \; H(f(x)) = f(-x).$$

F outputs real numbers, just evaluating the input function at 0, whereas G and H output functions: G increases the input function by 1 so that its graph moves up by 1 and H reflects the graph of a function in the y-axis (Figure 29(d)).

We cannot say whether F, G, H are continuous until we make clear which metrics we're using. But whether we use the maximum-distance-apart metric or the area-between metric, the functions G and H are continuous as the maximum distance between two graphs is unchanged by a vertical move of 1, nor is the area between the graphs. The same is equally true when H reflects the graphs in the y-axis.

Our choice of metric *does* matter for F, which evaluates an input $f(x)$ at $x = 0$. Continuity means being able to constrain the outputs by constraining the inputs: must two functions be close at 0 if the functions $f(x)$ and $g(x)$ are close? The answer is yes if we're using the maximum-distance-apart metric; the difference between the functions at the single input 0 can be no more than the greatest distance between the functions when considering all inputs. However, we cannot constrain the difference between the functions at 0 by constraining the area between the functions (Figure 29(e)). A function $g(x)$ with a tall, but very thin spike around $x = 0$ would produce a large difference between $f(0)$ and $g(0)$ whilst $f(x)$ and $g(x)$ would be close in terms of the area-between metric.

And we can now properly ask and answer the question that started this chapter: is the distance a car has travelled, $d(t)$, a continuous function of its speed, $s(t)$? (See Figure 22.) Any journey to work taking time T or less can be represented by a continuous function $d(t)$ on the interval $0 \leqslant t \leqslant T$. Is the function that takes input $s(t)$ to output $d(t)$ continuous? If two cars have speeds $s_1(t)$ and $s_2(t)$ that never differ by more than S then the two cars will never be more than distance ST apart on their journeys. So using maximum-distance-apart, we can constrain the distance between two journeys $d_1(t)$ and $d_2(t)$ by constraining the distance between their speeds $s_1(t)$ and $s_2(t)$. We have just shown $d(t)$ is a continuous function of $s(t)$ when using the maximum-distance-apart metric. Using a little calculus, a similar argument can be made for the area-between metric.

Equivalent metrics and continuity

Our main reason for introducing metric spaces was to provide a more general setting in which to discuss continuity. But sometimes *different* metrics on a set lead to precisely the *same* continuous functions. Two such metrics are the straight-line distance and taxicab distance on the plane.

To help appreciate this, we introduce the idea of **open balls** which generalize discs in the plane when we're using straight-line distance. The open ball $B(a, r)$ of radius r about a point a in a metric space is the set of points in the metric space that are less than distance r from a. When using straight-line distance in the plane, this is the interior of the circle of radius r with centre a (Figure 30(a)). The dashed circle in the figure indicates that the circumference is not included in the open ball.

However, if we were using the taxicab distance in the plane the ball $B(a, r)$ looks different. The taxicab distance between two points $a_0 = (x_0, y_0)$ and $a = (x, y)$ equals $|x - x_0| + |y - y_0|$ and so the points $a = (x, y)$ within (taxicab) distance r of a_0 are those satisfying

$$|x - x_0| + |y - y_0| < r.$$

Some algebraic manipulation shows that these points form the interior of a diamond (Figure 30(a)).

It is possible to fit such a diamond into any disc of the plane, and vice versa it is possible to fit a disc into any given diamond (Figure 30(b)). This is what it means for two metrics to be equivalent. More precisely, two different metrics d_1 and d_2 are said to be **equivalent** if any ball $B_1(a, r)$ contains some ball $B_2(a, s)$ and any ball $B_2(a, R)$ contains some ball $B_1(a, S)$. The subscripts here refer to which metric is being used and the smaller ball's radii s and S will usually be different from the radii r and R of the original balls.

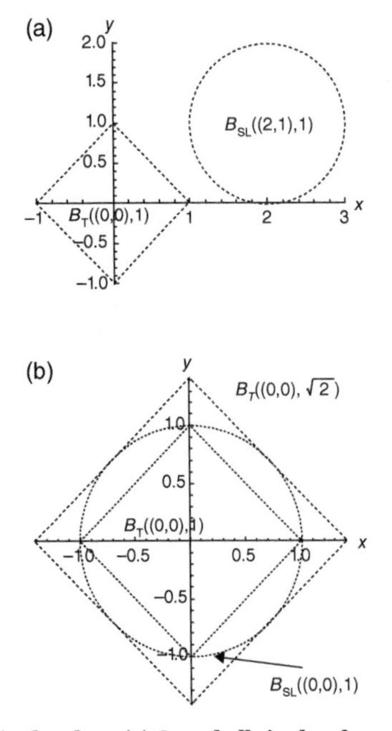

30. Open balls in the plane (a) Open balls in the plane, (b) Open balls within others.

The straight-line distance is always less than or equal to taxicab distance, as the former is distance measured 'as the crow flies'. But there are limits to how much greater taxicab distance can be compared with straight-line distance. The greatest taxicab distance from the centre of a circle of radius r to a point of the circle equals $\sqrt{2}r$. These are the points on the circle that are north-east, north-west, south-west, and south-east of the centre. This means that taxicab distance can be no more than $\sqrt{2}$ times bigger than straight-line distance, or expressed as an inequality,

$$d_{SL}(a,b) \leqslant d_T(a,b) \leqslant \sqrt{2}d_{SL}(a,b)$$

for any points a and b, and where the subscripts SL and T signify the metric. So $B_T(a, r)$ is contained in $B_{SL}(a, r)$ which is contained in $B_T(a, \sqrt{2}\, r)$, for any point a and radius r (Figure 30(b)).

By contrast, the maximum-distance-apart and area-between metrics for functions are not equivalent metrics. The maximum distance between two graphs *is* a constraint on how great the area between the graphs can be, but constraining the area between graphs does not constrain the maximum distance between two graphs. We saw this earlier in Figure 29(e) where two graphs differing only by a high, very thin spike have a large maximum distance between them but only a small area between them.

The important point here is that two different but equivalent metrics lead to the same functions being continuous. Now we can rewrite the definition of continuity in terms of open balls: a function $f(x)$, with inputs from M and outputs in N, is continuous at input $x = a$ if

for any positive e there is some positive d such that f sends $B(a, d)$ into $B(f(a), e)$.

This is purely a notational rewriting of the definition given on p. 69—all the points within distance d of a, namely those in $B(a, d)$, need to be sent to points within distance e of $f(a)$, namely sent into $B(f(a), e)$. If a function is continuous at $x = a$ when using a metric d_1 on M then it would also be continuous when using an equivalent metric d_2. This is because the ball $B_1(a, d)$ would contain a ball $B_2(a, D)$ using the second metric, and if $B_1(a, d)$ is sent into $B(f(a), e)$, then so is $B_2(a, D)$.

Open sets and continuity

An open ball is a basic example of a more general notion, that of an *open set*. Recall that an open ball consists of all points strictly less than a given distance away and so does not include the circumference of the ball. Similarly, open sets can be thought of as those sets that don't contain any of their boundary points. The formal definition is that a set U, contained in a metric space M, is **open** if around each point a in M there is an open ball $B(a,r)$ still contained in U (Figure 31(a)). Open balls are unsurprisingly examples of open sets. If a point b lies in $B(a, r)$, at a distance s from a, then the ball $B(b, r\text{-}s)$ is contained in $B(a, r)$ (Figure 31(b)). As before the dotted lines in Figure 31 denote that the boundary points aren't included in the sets.

Two equivalent metrics determine the same collection of open sets. To appreciate this, note that if a is in an open set U, we can place an open ball $B(a, r)$ around it inside U. If we have a second equivalent metric, then we can place another open ball centred at

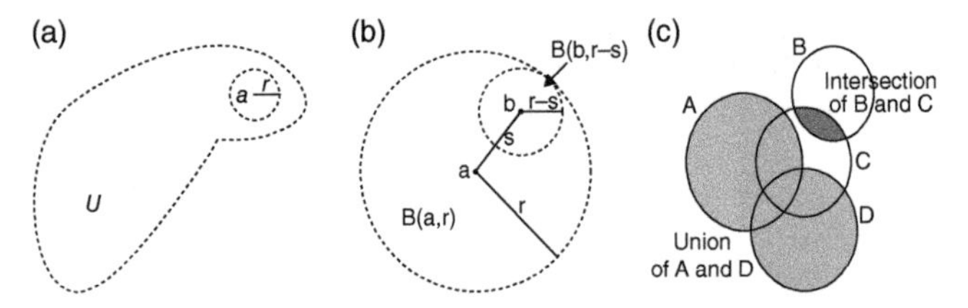

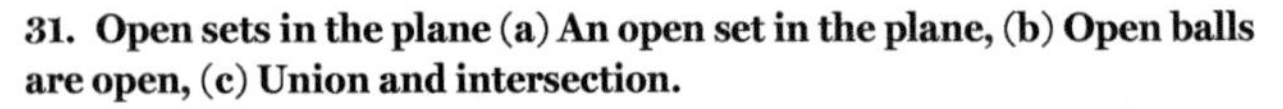
31. Open sets in the plane (a) An open set in the plane, (b) Open balls are open, (c) Union and intersection.

a—using this second metric—contained in $B(a, r)$ and so inside U as well. This means U is also an open set when using the second metric.

Equivalent metrics were mentioned earlier because a function that is continuous when using one metric is continuous when using an equivalent metric. And we have just seen that equivalent metrics lead to the same collection of open sets. These two facts are not unconnected and indeed *knowledge of which sets are open* is sufficient to determine which functions are continuous. It is relatively straightforward to show that the following definition of continuity in terms of open sets is equivalent to the ones given on p. 69 and p. 75; one advantage of this new definition is that it is much more generally applicable. Our new definition reads: a function f between metrics spaces M and N is continuous if

> whenever U is an open set in N, then the *preimage* $f^{-1}(U)$ is an open set in M.

The preimage $f^{-1}(U)$ consists of all elements of M that f sends into U. For example, if $f(x) = x^2 + 1$ and U is the interval $0 < x < 2$ then the pre-image $f^{-1}(U)$ is the interval $-1 < x < 1$ as these are precisely those x that satisfy the inequality $0 < x^2 + 1 < 2$.

That different metrics can lead to the same continuous functions and that continuity can be rephrased in terms of open sets, making no mention of metrics, suggest that the open sets, more so than metrics, are crucial to continuity. However, if we are going to take this approach—starting with open sets rather than metrics—we need to decide what properties open sets need to have; so far we have only defined open sets *in terms of* metrics. The two main properties of open sets are:

- for any collection of open sets, their *union* is also open;
- for any finite collection of open sets, their *intersection* is also open.

Given a collection of sets their **union** consists of those elements that are in one or more of those sets; the **intersection** of those sets consists of those elements that are in every one of those sets (see Figure 31(c)). The union of *any* collection of open sets is open, but in general the intersection of an *infinite* collection of open sets need not be open. Consider the interval $-r < x < r$ where r is a positive number. If we consider all such sets for positive r, then the only element in each set is 0. So the intersection of these sets is $\{0\}$, the set with sole element 0, an intersection which is not open.

Felix Hausdorff sought to capture and generalize the necessary properties of open sets in his 1914 magnum opus *Grundzüge der Mengenlehre* or *Essentials of Set Theory*. There Hausdorff introduced many key ideas of set theory and general topology, extending and synthesizing the work of Fréchet and others into a coherent whole. It was also here that he coined the term metric space for the spaces Fréchet had introduced.

Given a *set* M, then any collection $\mathcal{T}$ of sets in M which satisfies the following:

- for any collection of sets in $\mathcal{T}$, then their union is also in $\mathcal{T}$;
- for any finite collection sets in $\mathcal{T}$, then their intersection is also in $\mathcal{T}$;
- M is in $\mathcal{T}$;
- the empty set—the set with no elements—is in $\mathcal{T}$.

is called a **topology** on M and a set M, with a topology of sets $\mathcal{T}$, is known as a **topological space**.

Note that M, by itself, is 'just' a set, not a metric space, with elements having no sense of being close to one another; the collection of sets $\mathcal{T}$ is an effort to separate out M without necessarily going so far as introducing a metric. Given a metric space then its open sets form a topology, but not all topologies arise from metrics.

The following is an example of a topology on the set of whole numbers which does not arise from a metric. Necessarily the entire set of whole numbers and the empty set must be in the topology; the only other sets in this topology are those of the form

$$U_n = \{n, n+1, n+2, n+3 \ldots\},$$

where n is a whole number. The intersection of finitely many U_i is $U_{\max(i)}$, where $\max(i)$ denotes the largest of the is, and so in the topology; the union of any collection of U_i is $U_{\min(i)}$ if the minimum $\min(i)$ exists and is otherwise the set of all whole numbers—in either case the union is in the topology. So, we have just verified that this collection of sets satisfies the rules of being a topology. In this space non-empty open sets always overlap, which isn't true of metric spaces, so this is one way to be sure this topology doesn't arise from a metric.

The smallest topology $\mathcal{T}$ on M just includes the empty set and the set M. This is known as the *trivial* topology—it does not separate out the elements of M at all and the only continuous functions on M are the constant ones. The largest topology is the *discrete* topology in which case every set is in $\mathcal{T}$, including single points; it introduces extreme separation, placing each element of the set away from others and *all* functions on M are continuous. In practice, important topologies are somewhere between these two extremes. Many important topologies arise from metrics, but there are important ones that do not, such as the *Zariski topology*, important in algebraic geometry which studies sets defined by polynomial equations.

Working with topological spaces, rather than metric spaces, does more than just generalize further the ideas of continuity. Many proofs in topology appear cleaner, with the logic of the proof and the rules of being a topological space meshing together much more naturally.

Convergence and continuity

Metric spaces are also a natural setting in which to define *convergence*, a central idea of calculus and of mathematical analysis. A **sequence** from a set is a list of elements in that set; for example, the following are three sequences of real numbers.

$$1, \frac{1}{2}, \frac{1}{3}, \frac{1}{4}, \frac{1}{5}, \ldots \qquad -1, 1, -1, 1, -1, \ldots \qquad 1, 2, 3, 4, 5, \ldots$$

The ellipsis '…' at the end of each list denotes that the list goes on forever. A sequence is commonly denoted as $x_1, x_2, x_3, \ldots$ and the terms in these sequences can each be described by giving a formula for the nth term x_n:

$$x_n = \frac{1}{n}, \qquad x_n = (-1)^n, \qquad x_n = n.$$

Only the first of these sequences converges; the terms of the sequence are steadily getting smaller and I hope it's not surprising that this sequence converges to 0 which is known as the sequence's **limit**. The second sequence does not converge to any limit; some of its terms though—like the even terms (second, fourth, sixth, etc.) which are all 1—*do* converge but overall the sequence does not. The final sequence does not converge either and in fact no selection of its terms converges either.

The very word 'converge' suggests that the terms of the sequence get closer and closer to some limit. Rigorously, a sequence x_n, in a metric space M, **converges** to a limit a in M if any ball $B(a, r)$ contains a tail of the sequence; this means that from some term onwards, all remaining terms of the sequence are inside $B(a, r)$.

Using this definition, we can see why the sequence $x_n = 1/n$ converges to 0. The ball $B(0, r)$ contains all terms x_n of the sequence where $n > 1/r$, which is a tail of the sequence. In Figure 32(a), a sequence spirals in to its limit (0, 0). The pictured

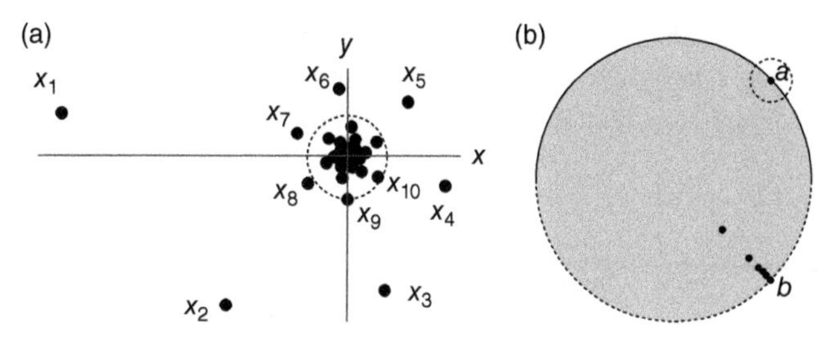

32. Visualizing convergence (a) A sequence converging in the plane, (b) Neither open nor closed set.

dashed disc $B((0, 0), r)$ contains x_{10} and every term afterwards; we would need to go farther down the sequence to find a tail contained within a smaller open ball. If a sequence converges to a limit in a metric space, then the sequence converges to the same limit if we use any equivalent metric.

Sequences also provide an alternative means of defining continuity. For a continuous function f between metrics spaces M and N, if a sequence x_n of inputs converges to a limit a in M, then the sequence $f(x_n)$ of outputs converges to $f(a)$. For example, with the sequence $x_n = 1/n$ which converges to 0, and the continuous function $f(x) = x^2$ then $f(x_n)$ is the sequence $1, \frac{1}{4}, \frac{1}{9}, \frac{1}{16}, \frac{1}{25}, \ldots$ which converges to $0 = 0^2 = f(0)$. Similarly for the continuous function $g(x) = 2x + 1$, then $g(x_n)$ is the sequence $3, 2, 1\frac{2}{3}, 1\frac{1}{2}, 1\frac{2}{5}, 1\frac{1}{3}, \ldots$ which converges to $1 = 2 \times 0 + 1 = g(0)$. And the converse is also true: if, whenever a sequence x_n converges to a, then $f(x_n)$ converges to $f(a)$, the function $f(x)$ is continuous at $x = a$.

Having defined convergence, we can now define what it means for a set to be closed. If C is a set in a metric space M, then it is **closed** if whenever a sequence of points in C converges then the limit is *also* in C. Closed sets might be thought of as those sets that contain *all* their boundary points. An alternative definition of C being a closed set in M is that the *complement* of C, that is all the points in M that aren't in C, is an open set.

So, an open set is one which contains *none* of its boundary points, and a closed set is one which contains *all* its boundary points. Clearly most sets fall into neither of these categories and will contain *some but not all* of their boundary points; it's important to realize that being open and being closed are not opposites of one another. In Figure 32(b) is a set which is neither open nor closed: a disc with its upper circumference included but the lower half omitted. The boundary point a is contained within the set so the set is not open—or equally no open ball $B(a, r)$ around a is contained within the set. And the boundary point b isn't contained within the set, so the set is not closed—or equally there is a sequence of points *in the set* that converges to b despite b not being in the set. And some sets can be *both* open and closed, for example the whole plane is both an open and closed set of the plane—there are *no* boundary points, so all and none of them are simultaneously in the set.

Subspaces

Any set S in a metric space M inherits the structure of a metric space, as two points of S are also points of M and we may assign them the same distance as before when they were considered as points of M. For example, if M was the set of cities in the United States, and S is the set of Californian cities, then it's natural to think of Los Angeles and San Francisco as being the same distance apart whether they're being considered as American cities or Californian cities. The set S naturally becomes a metric space, in its own right, called a **subspace** of M.

Subtleties arise when we consider the open sets of and continuous functions on a subspace. It may help to remember the Flatland mindset mentioned in Chapter 1; we need to imagine life as an inhabitant of the subspace S, as if Californians somehow cannot see the rest of the USA.

For example, let M be the real line and let S be the union of the intervals $0 \leqslant x \leqslant 2$ and $3 < x \leqslant 4$ (Figure 33(a)). What are the open balls $B(1, 1)$ and $B(2, 1)$ as open balls in S? By definition, $B(1, 1)$ is the set of points in S within distance 1 of the point 1; this is the interval $0 < x < 2$ which agrees with what $B(1, 1)$ is in the real line M. However $B(2, 1)$ is the interval $1 < x \leqslant 2$ as an open ball of S; this is the set of points in S that are within distance 1 of 2. This is different from the situation with the real line, and is yet more surprising when we remember open balls are open sets. The point 2 appears incongruous as points near it to the right are not included but, from the Flatland mindset of the subspace S, those points are essentially invisible as they are not in S. So, the interval $1 < x \leqslant 2$ is indeed an open set in S.

Also, perhaps surprisingly, the function

$$f(x) = \begin{cases} -1 & \text{for } 0 \leqslant x \leqslant 2 \\ 1 & \text{for } 3 < x \leqslant 4 \end{cases}$$

is continuous on the subspace S (Figure 33(b)). Instinctively you notice a jump between the outputs of –1 and 1 and may think f is not continuous, but there is no point in S where a discontinuity can be identified.

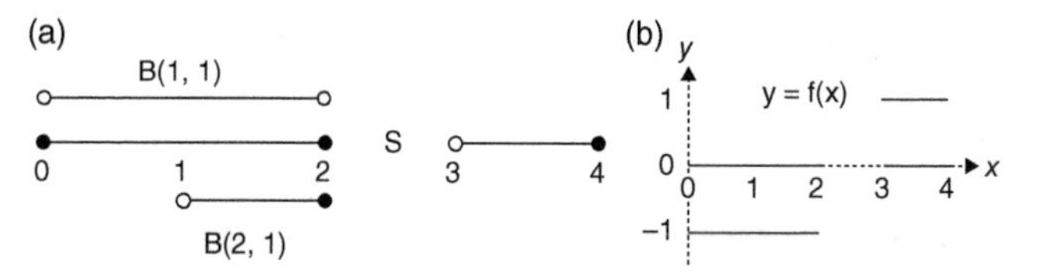

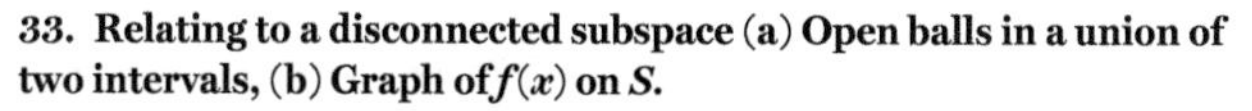
33. Relating to a disconnected subspace (a) Open balls in a union of two intervals, (b) Graph of $f(x)$ on S.

Note also that the interval $0 \leqslant x \leqslant 2$ is open as a set in S. This interval is the open ball $B(1, 2)$ in S and open balls are open sets; those points in S which are within distance 2 of the point 1 form the interval $0 \leqslant x \leqslant 2$. Similarly, the interval $3 < x \leqslant 4$ is open as a set in S as it is the open ball $B(4, 1)$ in S. As the complement of an open set is a closed set, then both intervals are also closed sets in S. This again may seem counter-intuitive as 3 looks to be a boundary point that's missing from the interval $3 < x \leqslant 4$. But from the Flatland mindset of S the point 3 is effectively invisible and so not missing at all.

Compactness and connectedness

In Chapter 3 we met two important theorems relating to continuous functions on an interval $a \leqslant x \leqslant b$, namely the *boundedness theorem* and the *intermediate value theorem*. In general continuous functions need not be bounded—for example x^2 on the real line is unbounded—nor need they attain intermediate values—for example the function $f(x)$ in Figure 33(b) does not take the value 0 despite attaining −1 and 1. There must be some property or properties of the interval $a \leqslant x \leqslant b$ of inputs that means continuous functions *on that domain* are bounded or attain intermediate values.

One of these properties can be captured in terms of the behaviour of sequences in such an interval. A sequence like −1, 1, −1, 1, −1,... may not converge, but some selections of the sequence do—for example the odd terms (first, third, fifth...) are all −1 and so converge to −1 and the even terms (second, fourth, sixth...) similarly converge to 1. But some sequences like 1, 2, 3, 4, 5,... have no selection that converges; any selection grows towards infinity. The **Bolzano–Weierstrass theorem** states that any sequence in an interval $a \leqslant x \leqslant b$ has a selection which converges *in that interval*.

A metric space in which the Bolzano–Weierstrass theorem holds, so that all sequences in the space have a selection that converges to a limit in the space, is called **compact.** The theorem shows the interval $a \leqslant x \leqslant b$ is compact, the sequence 1, 2, 3, 4, 5 . . . in the real line shows that the real line is *not* compact as no selection converges, and the sequence $1, \frac{1}{2}, \frac{1}{3}, \frac{1}{4}, \frac{1}{5} \ldots$ shows the interval $0 < x \leqslant 1$ is not compact as it has no selection that converges *in that interval*; this last sequence's limit of 0 is outside the interval and any selection from this sequence also converges to 0.

For the real line, plane, and higher dimensional equivalents, the **Heine–Borel theorem** states that the compact sets are precisely the closed and bounded sets. We have already defined what a closed set is, and a set is *bounded* if it is contained in some ball $B(0, R)$. So we can see that the real line is not compact as it is not bounded and the interval $0 < x \leqslant 1$ is not compact as it is not closed, missing the boundary point of 0.

Compactness was first introduced by Fréchet in 1906, and the compactness of $a \leqslant x \leqslant b$ is the reason that the boundedness theorem holds. More generally it is true that real-output continuous functions on a compact domain are bounded and attain their bounds. It is also true that compactness is a topological invariant, so that two homeomorphic metric spaces are either both compact or neither is compact. Compactness is a much more important notion in topology than these few examples can make clear, with many theorems of topology concerning compact spaces.

However, the intermediate value theorem on the interval $a \leqslant x \leqslant b$ is not a consequence of compactness, but rather of *connectedness*, a notion first introduced by Hausdorff in 1914. Intuitively, connectedness is a simple idea: being connected means being 'in one piece', but how can we capture this definition? Surely the real line is connected and the set of whole numbers disconnected, as ought to be the earlier space S which is the union of the intervals $0 \leqslant x \leqslant 2$ and $3 < x \leqslant 4$.

Intuition tells us that S is disconnected and, more, that S is made up of just two pieces, the two intervals. We noted earlier that the interval $0 \leqslant x \leqslant 2$ is both open and closed as a set in S, as is the second interval. In general, for a metric space M, the whole set M and the empty set—the set with no points of M—are both open and closed sets in M. A space M is defined to be **connected** if these are the *only* open and closed sets in M. Or equally a space is connected if it cannot be split as the union of two non-empty, open sets with no elements in common.

The first interval $0 \leqslant x \leqslant 2$ and second $3 < x \leqslant 4$ which make up the set S are called the *connected components* of S; these are the largest sets contained in S that are themselves connected. For the set of whole numbers, the connected components are the sets of individual numbers such as $\{1\}$ or $\{3\}$. For a space to have just one connected component is just another way to say that a space is connected.

The intermediate value theorem holds on a connected space: let M be a connected metric space, f be a real-valued continuous function on M and a,b be points in M such that $f(a) < 0 < f(b)$. Then there is a point c in M such that $f(c) = 0$.

This definition of connectedness may seem rather abstract, and there is a more concrete version of connectedness, **path-connectedness**, which is easier to appreciate. Given points a and b in a metric space M, then a *path* between a and b is a continuous function p from the interval $0 \leqslant x \leqslant 1$ to M such that $p(0) = a$ and $p(1) = b$ (Figures 34(a), 34(b)). You might consider p as being a journey from the point a as a starting point when $x = 0$ to the point b as finishing point when $x = 1$. The points $p(x)$ where $0 < x < 1$ are the points passed through getting from a to b.

Path-connected metric spaces are connected, though there are some weird spaces that are connected without being path-connected (Figure 34(c)). For many regions, paths between points can be straightforwardly defined and a space quickly shown to be

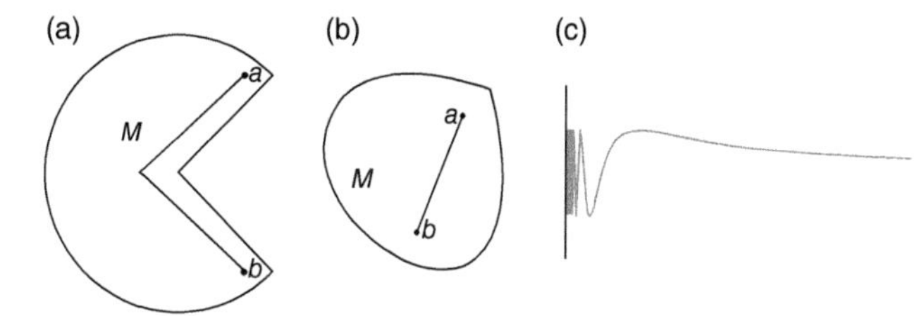

34. Relating to path-connectedness (a) A path-connected set, (b) A convex set, (c) Topologist's sine curve.

path-connected and so connected. For example, in an open disc, the plane, a half-plane, any two points are connected by a straight line. These are all examples of *convex sets*, which means that given any two points in the set then the line segment between them is also contained in the set (Figure 34(b)) which is not true of the Pacman-like shape (Figure 34(a)).

The *topologist's sine curve* in Figure 34(c) is an example of a space which is connected but not path-connected. The space is the union of the y-axis and the curve $y = \sin(1/x)$ for $x > 0$. Loosely put, the space is connected because the sine curve gets arbitrarily close to the y-axis but it is not path-connected as the function $\sin(1/x)$ has no limit as x becomes small. Any two points on the sine curve can be connected by a path, as can any two points on the y-axis, but a point on the y-axis and a point on the sine curve cannot be connected by a path in the space.

Again, connectedness is a topological invariant—if two metric spaces are homeomorphic and one is connected then so is the other. Similarly, path-connectedness is a topological invariant.

Topological invariants

In Chapter 1 we separated out which letters of the alphabet were topologically the same—homeomorphic—or not. For two

Table 5. Eight spaces which are not homeomorphic

Compact	A closed disc	A figure 8	The interval $0 \leqslant x \leqslant 1$	A circle
Not compact	A closed half-plane	A plane	The interval $0 < x \leqslant 1$	A line

equivalent letters, we described a way of deforming each into the other; for topologically different ones, we needed to find a feature—a topological invariant, like the T-junction in the E—that would remain a feature of the letter, even when deformed. Compactness and connectedness are topological invariants and so can be used to differentiate between spaces.

In Table 5 are eight spaces, none of which is homeomorphic to another. Some are compact—closed and bounded—others not, so compactness can be used somewhat to separate them. Using the Heine–Borel theorem, we see the top row are compact, each being closed and bounded, and the bottom row are not. So none of the top row is homeomorphic to one from the bottom row, but compactness implies nothing about whether spaces on the same row are topologically different. (A closed disc means a circular disc including its circumference. A closed half-plane means all the points on and to one side of a line in the plane; so the bounding line is included in the closed half-plane.)

All the spaces are connected, but with a little imagination we can still make use of connectedness to discern topological differences. A point of a connected space which, when removed, disconnects the space is called a **cut point**. We also noted in Chapter 1 that the T-junction in an E is the special cut point which, if removed, breaks E into three connected components; removing a different point leaves just two components remaining.

Looking at the top row of Table 5, neither the circle nor the closed disc has any cut points, but the figure 8 and interval do have; this means the former two are not homeomorphic to the latter two. However, we can disconnect the circle by removing *two points* and such is not true of the closed disc. Finally, the middle point of the figure 8 is the *only* cut point whilst any point of the interval, except the ends, is a cut point.

In the bottom row, the line and interval have cut points, but not the half-plane nor plane. But 1 is not a cut point of the interval whilst every point of the line is a cut point. All that remains to do is show the closed half-plane and plane aren't homeomorphic.

For this we need a further topological invariant, discussed in more detail in Chapter 5. The boundary points of a closed half-plane seem different from other points. Neither the plane nor half-plane has any cut points, but removing a boundary point from the half-plane doesn't make a hole, the way removing a point not on the boundary does. Informally put, a space with no holes is called *simply connected* and this is a topological invariant. As the removal of a point of the half-plane can leave a simply connected remainder (without holes), and as this is not true of the plane, then the closed half-plane and plane are not homeomorphic.

Taking the methods and theorems of continuity for single input, single output functions and applying them to the more general settings of metric and topological spaces proves to be a very powerful approach. Consequently, properties such as compactness and connectedness are used widely in mathematics and a classical theorem guaranteeing a real number solution to an equation might in a modern setting show that there is a continuous function which solves a differential equation. Seemingly these are very different mathematical problems, but this abstract mode of thinking helps mathematical ideas be applied in their fullest generality.

Chapter 5
Flavours of topology

From the mid-19th century, topological understanding progressed on various fronts. The *geometric topology* of Chapter 2 concerned surfaces and grew out of the work of Euler, Möbius, Riemann, and others. The *general topology* of Chapters 3 and 4 was more analytical and foundational in nature; Hausdorff was its most significant progenitor and its growth mirrored other fundamental work being done in set theory. There are yet more flavours of topology—a topologist might have texts on their bookshelves entitled *algebraic topology, differential topology, symplectic topology,* and other books on the subject with less explicitly topological titles.

In Chapter 2 we met the Euler number and saw that it, together with knowing whether a surface is one- or two-sided, identifies the shape of a closed surface. Further, the Euler number imposes *global constraints* about what is possible on a closed surface as we'll see in the next two sections.

Differential topology

The continuous functions form an important class of functions, but we also saw in Chapter 3 that continuous functions can still be quite nasty; for example, the blancmange function does not have a defined gradient at *any* point. The *smooth functions* also form an important class. A function might be continuous but still fail to be smooth by changing in a jerky fashion; smooth functions by contrast have a defined gradient everywhere. For example, the speed function $s(t)$ in Figure 22(b) is continuous but doesn't have a defined gradient at the times $t_1, \ldots t_6$. Before and after those times, the gradient is clear but there is no defined gradient of $s(t)$ when the car accelerates or brakes suddenly. The study of functions that change smoothly on surfaces lies within the field of **differential topology** and we will see that the Euler number of a closed surface impacts on what properties smooth functions on a closed surface can (or must) have.

An important first result about smooth functions is that the gradient of the function is zero at any maximum or minimum. Note, in Figure 24(a), how the gradient of $\sin x$ is zero at its extreme values whereas, in Figure 27(a), the (continuous but not smooth) function $f_1(x)$ has no defined gradient at its maxima. This result is known as *Fermat's theorem* and these maxima and minima are known as **critical values** and the corresponding inputs as **critical points**.

But a function of two inputs can have a greater variety of critical points than a function with just one input. When we have two inputs and one output, the notion of gradient is a little less clear. Sketched in Figure 35 are three surfaces $z = f(x, y)$ exhibiting different types of critical point.

In Figure 35(a) we have a *minimum* at the bottom of a bowl-like graph, above the point $x = 0, y = 0$. In whatever direction we move from the bowl's bottom, we move upwards. Likewise, in Figure 35(b), we have a hill-like graph with a *maximum* at the top and however we move away from that maximum we move downwards. The point $x = 0, y = 0$ in Figure 35(c) is neither a maximum nor a minimum. The function is zero at that point and if we move along the x-axis we get to a point where $f(x, 0) = x^2 > 0$ so that the function has increased; but if we were to move along

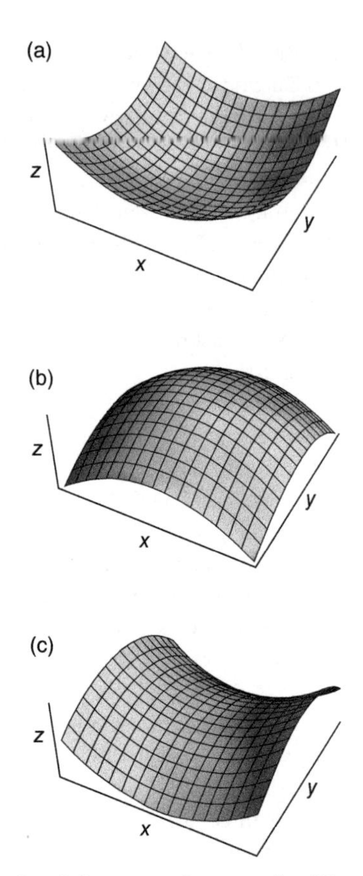

35. Examples of a minimum, maximum, and saddle point (a) $f(x,y) = x^2 + y^2$, **(b)** $f(x,y) = -x^2 - y^2$, **(c)** $f(x,y) = x^2 - y^2$,

the y-axis we get to a point where $f(0, y) = -y^2 < 0$ so that the function has decreased. This critical point is an example of a *saddle point* and hopefully the name is not surprising given the shape of the graph. As with a saddle on a horse's back, the arc of the horse's back has a lowest point where the rider sits, but the rider straddles the horse in a manner where the saddle is at the

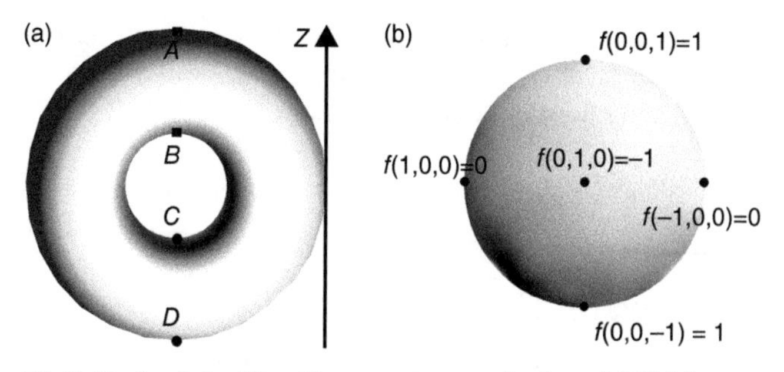

36. Critical points of functions on a torus and sphere (a) Height function on a torus, (b) $z^2 - y^2$ on the unit sphere.

highest point. There are yet more complicated examples of critical values with two inputs; the function $f(x, y) = x^3 - 3xy^2$, for example, has a graph with a saddle-like critical point that would suit a rider with two legs and a tail—so this is called a *monkey saddle*—but in what follows we will be interested in functions that *only* have maxima, minima, and saddle points as critical points.

Consider now the height function z for the torus drawn in Figure 36(a). The height function has a maximum value at the top of the torus A, and a minimum value achieved at the bottom of the torus D. In fact, as the torus is compact, the boundedness theorem guarantees that any continuous function on it must have at least one maximum and at least one minimum.

There are two further critical points: C, at the bottom of the hole and B, at the hole's top. If we move through C, passing through the hole, then our height is at its greatest as we pass through C; but if we slide down one side of the hole and up the other side, then the lowest point of our journey is at C. This critical point C is a saddle point and similar paths can be made through B which include it as the highest or lowest point on the journey, so B is also a saddle point.

In Figure 36(b), we consider the function $f(x, y, z) = z^2 - y^2$ on the sphere with equation $x^2 + y^2 + z^2 = 1$. With a little work we can show that this function's critical points are

- maxima at (0, 0, 1) and (0, 0, –1), the sphere's north and south poles;
- minima at (0, 1, 0) and (0, –1, 0), two points lying on the sphere's equator;
- saddle points at (1, 0, 0) and (–1, 0, 0), two more points lying on the sphere's equator.

At the north pole (0, 0, 1), z is as large as it can be on the sphere, so any move away means that $z^2 - y^2$ decreases and (0, 0, 1) is therefore a maximum of f. The same argument applies at the south pole. Similarly, at (0, 1, 0), y is as large as it can be, so any move away means that $z^2 - y^2$ increases and so (0, 1, 0) is a minimum, as is (0, –1, 0). Finally, at (1, 0, 0), x is as large as it can be. As we move from (1, 0, 0) then x will decrease and either y or z or both might increase, meaning that f may decrease or increase. This means that (1, 0, 0), and likewise (–1, 0, 0) are saddle points of f.

At this point, any relationship with topology probably seems unclear. The height function on the torus had one maximum, one minimum, and two saddle points; $z^2 - y^2$ on the unit sphere had two maxima, two minima, and two saddle points. However, if we continued considering smooth functions on the torus, in each case we would find that

$$\text{number of maxima} + \text{number of minima} - \text{number of saddles} = 0$$

(as with the height function where $1 + 1 - 2 = 0$) and for any smooth function on the sphere we'd find that

$$\text{number of maxima} + \text{number of minima} - \text{number of saddles} = 2$$

(as with $z^2 - y^2$ where $2 + 2 - 2 = 2$). That itself is perhaps surprising enough, but when we remember that 0 and 2 are the Euler numbers of the torus and sphere then we might suspect a deeper relation. Generally, then, it is true for any smooth function on a closed surface that

$$\text{number of maxima} + \text{number of minima} - \text{number of saddles} = \text{Euler number of the surface.}$$

In this way the Euler number is a *global constraint* as to what features smooth functions may have on a closed surface. We might tweak our function near any point to create a maximum, a minimum, or a saddle point, but there will necessarily be consequences for the function elsewhere. The Euler number limits overall possibilities in the unavoidable way of if-you-push-down-here-it-pops-up-over-there DIY problems. As an example, we can deduce that any smooth function on a torus *must* have at least two saddle points: a smooth function on a torus must have at least one maximum and at least one minimum because the torus is compact, and so

$$\text{number of saddles} = \text{number of maxima} + \text{number of minima} \geqslant 1 + 1 = 2.$$

In fact, we can see that the sphere (being the only closed surface with an Euler number of 2 or more) is the only closed surface on which a smooth function may have no saddle points. An example of such a function is the height function on a sphere.

The Scottish physicist James Clerk Maxwell was one of the first to appreciate such a relation amongst critical points in an 1870 paper *On Hills and Dales*, but he was working solely with functions on the plane—and so arrived at the number 1 (the Euler number of the plane or a punctured sphere). Poincaré would generalize the result to closed surfaces, using techniques that would now be considered *Morse theory*, after Marston Morse who, from 1925 onwards, would prove a series of deep results connecting the topology of spaces and the analysis of functions on those spaces.

The hairy ball theorem

Informally put, the **hairy ball theorem** says that you can't comb a hairy ball flat without creating a tuft or cow-lick where the hair refuses to lie flat. We might brush and sweep the hair on the sphere in various styles but something about the sphere's underlying shape means we can't ever get the hair to lie flat everywhere. But it's not hard to imagine how the hair on a torus could all be swept flat in the same direction (Figure 37(b)). As before, with smooth functions, we'll see that it's the sphere's topology that makes this impossible.

The hair in this theorem is a metaphor for a *tangent vector field* which might be easiest thought of as a fluid flowing on a surface; the hairy ball theorem then states that for any fluid flow on a sphere, there will be one or more points where the fluid is still and unmoving, the way in the eye of a storm the wind is calm (Figure 37(a)). Such points, where the fluid is still, are called **singularities** and there are various types of singularity that a

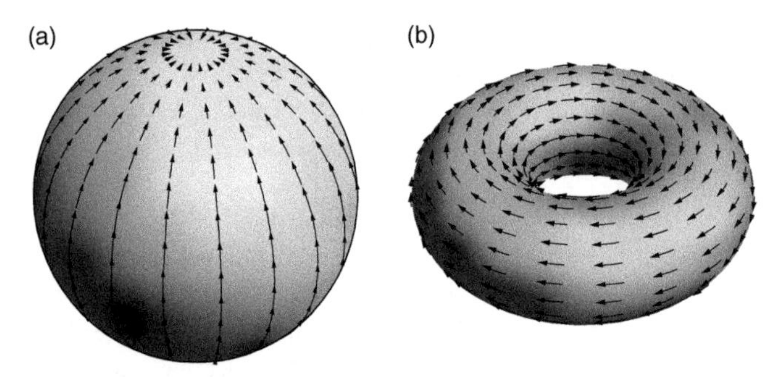

37. Vector fields on the sphere and torus (a) A hairy ball with no hair at poles, (b) A hairy-everywhere torus.

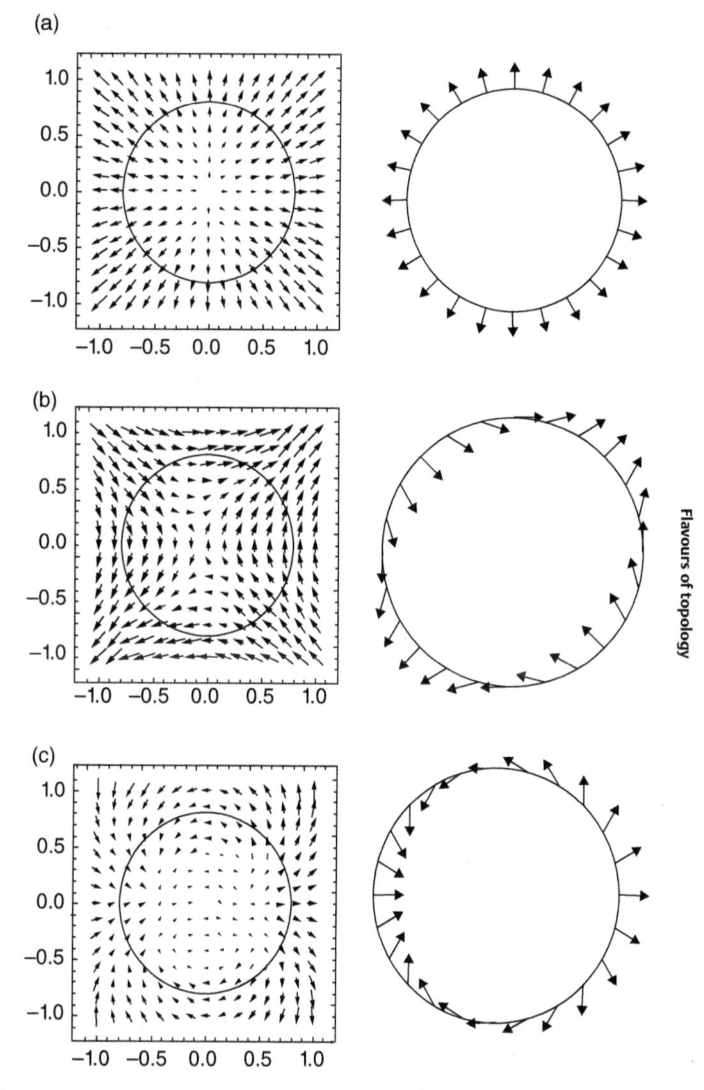

38. Examples calculating indices of vector fields (a) $v(x, y) = (x, y)$, (b) $v(x, y) = (y, x)$, (c) $v(x, y) = (x^2 - y^2, 2xy)$.

vector field can have. Three vector fields in the plane are shown in Figure 38, with the origin (0, 0) being the singular point for each.

The velocity of the flow at the point (x, y) is given by a formula $v(x, y)$. Figure 38(a) has a *source* at (0, 0) with the fluid coming out of the origin; Figures 38(b) and 38(c) are more complicated examples with fluid flowing in and out of the origin along different lines.

We can associate with a singularity a whole number called its **index**. Around each of the singularities in Figure 38 is drawn a circle, redrawn to the right of each figure with the vector field showing on the circle as arrows. If we walk anti-clockwise around the circle then the arrows initially point in a certain direction, change as we move around the circle, but ultimately must return to their initial direction as we complete the circle. The index of the singularity is the number of times that the arrows themselves have gone around anti-clockwise.

So, in Figure 38(a) if we start at 12 o'clock the arrow points north and as we move anti-clockwise the arrows move around to west at 9 o'clock, south at 6 o'clock, east at 3 o'clock, and ultimately back to north. The arrows themselves have gone once anti-clockwise around the compass (north–west–south–east–north) and so the index is 1 in this case. In Figure 38(b) at 12 o'clock the arrows point east; as we move clockwise around the circle they go from east to south (at 9 o'clock) to west to north to east—this journey of east–south–west–north–east is a single journey once around the compass *clockwise* and so the index in this case is −1. Finally Figure 38(c): at 12 o'clock the arrow is pointing west, by 10.30 it's pointing south, by 9 o'clock it's pointing east, by 6 o'clock it's already back to west, and by 12 o'clock the arrows have gone around one further time anti-clockwise before returning to west. The arrows have been around twice in an anti-clockwise fashion and the index in this case is 2.

Poincaré's theorem is a generalization of the hairy ball theorem and states that

> *The indices of the singularities of a vector field on a closed surface add up to its Euler number.*

(Indices is the plural of index.) If we look at the flow in Figure 37(a) then we see that there is a source at the north pole which has index 1, and a sink at the south pole which a check shows also has index 1 and then 1 + 1 = 2 adds up to the Euler number of the sphere. The hairy-everywhere torus in Figure 37(b) has no singularities, so the sum of the indices is 0, the Euler number of the torus.

The hairy ball theorem is then a consequence of Poincaré's theorem—if there were a flow on the sphere with no singularities at all, then the Euler number of the sphere would be 0 and we know it is 2. In fact, we can see that the only closed surfaces on which there may be flows without singularities are the torus and Klein bottle, those being the only ones with an Euler number of 0. Again, we see the Euler number is a global constraint as to what flows are possible on a surface. We might at any point stir a fluid in such a way as to make a singularity of our choosing, but there will necessarily be consequences elsewhere on the surface, so that the total sum of the indices still equals the Euler number. Poincaré proved his theorem in 1881, which was later generalized in 1926 by Heinz Hopf to higher-dimensional equivalents of surfaces (manifolds). Consequently, the theorem is often called the *Poincaré–Hopf theorem.*

In Chapter 2, we classified surfaces up to continuous equivalence (homeomorphism)—we might instead have considered smooth surfaces up to smooth equivalence (the technical word being *diffeomorphism*). At first glance these are seemingly different classification problems—going back to the alphabet I hope you can see that C and I can be smoothly deformed into one another, but neither into a V because of its pointy base. Continuous deformations can be jerky but smooth ones need to be fluid. However it turns out in two dimensions that the classification problem has the same solution: if we consider a *smooth* closed surface it must be diffeomorphic to (a smooth version of) precisely one of $\mathbb{T}^{\#g}$ or $\mathbb{P}^{\#k}$.

The classification theory for three-dimensional manifolds is yet more complicated and its resolution brings us into the 21st century, but it remains the case that the continuous and smooth versions of the classification problem lead to the same solution. But the four-dimensional problem provided further colour in the 1980s, when it was found that it is impossible to 'make smooth' some topological four-dimensional manifolds and some others are 'smoothable' in essentially different ways.

Structures

Shortly we will move on to a discussion of *algebraic topology* which associates with a surface (and other spaces) algebraic structures that capture something of the essence of the surface's topology. In Chapter 4 we met compactness and connectedness, and these are important topological invariants but, ultimately, they are also *binary*. A space can be compact or not—there are no shades of grey here. It would be useful to have an invariant that retained something subtler of the topological character of a space. Before that we will need to consider the algebraic structures such topological invariants might take the form of.

Much of mathematics is concerned with the study of structures. The objects of mathematics—whether numbers, sets, functions—are often in some relation to one another or may be combined in certain ways. Much of the technical language of mathematics describes the details governing such relationships and combinations. A set (loosely speaking) is a collection of objects without any further structure and we needed, for example, a metric to introduce a notion of distance to a set to make it a metric space. The important sets of mathematics typically come naturally with some further structure. For example, with the natural numbers

$$\{0, 1, 2, 3, 4 \ldots\},$$

we might identify 0 as the smallest of these numbers, but then we're already recognizing the set's implicit order, and we might recognize the numbers can be added together but not subtracted, at least not if we want the result to remain in the set (e.g. $3 + 4 = 7$ is in the set but $3 - 4 = -1$ is not).

One of the most common algebraic structures within mathematics is a **group**. A group is a set G together with an operation $*$ which, for two inputs x and y from G, combines them into an output which we denote as $x * y$ and which is importantly also in G. The set G might be a set of numbers, functions, geometric transformations or even a set of sets, and the operation $*$ might be addition, multiplication, composition of functions or some set operation. But, to be a group, further rules must also apply: for G and $*$ to form a group we need:

1. $x * (y * z) = (x * y) * z$ for all x, y, z in G;
2. there is an element e in G such that $e * x = x = x * e$ for all x in G;
3. for any x in G there is an element x^{-1} in G such that

$$x * x^{-1} = e = x^{-1} * x.$$

The element e in rule 2 is called the *identity* and the element x^{-1} which combines with x to give e is called the *inverse* of x. Rule 1 is known as *associativity* and the purpose of this rule is that any product like $a * b * c * d$ needs no further clarification—we'll always arrive at the same answer no matter in what order we carry out the three multiplications involved in this product.

The natural numbers and addition meet rules 1 and 2 (with $e = 0$) as

$$x+(y+z)=(x+y)+z \quad \text{and} \quad 0+x=x=x+0$$

for any natural numbers x, y, and z, but fail rule 3, as there is no natural number x^{-1} we can add to $x = 2$ to get a sum of $e = 0$. We would like x^{-1} to be -2 but this is not a natural number, being negative. However, there are many important sets and operations that do meet the three rules—the subject *group theory* is a significant part of modern algebra—but as our focus here is topology, we only mention a handful:

- The set of whole numbers, denoted $\mathbb{Z}$, with operation +. In this case $e = 0$ and $x + (-x) = 0$.
- The set of non-zero real numbers, denoted $\mathbb{R}^*$, with operation $\times$. In this case $e = 1$ and $x \times (1/x) = 1$.
- The set of rotations in the plane about the origin, with the operation $\circ$ denoting composition. So $x \circ y$ means doing rotation y and then doing rotation x. Here e is rotation through zero degrees and if x is a rotation through some angle anti-clockwise, then its inverse x^{-1} is the rotation through the same angle clockwise.
- The set $\mathbb{Z}^2$ of pairs (x, y) of whole numbers with the operation + defined by

$$(x,y)+(m,n)=(x+m,y+n)$$
$$\text{and}\, e=(0,0),\ \text{noting that}\,(x,y)+(-x,-y)=(0,0).$$

If we could assign groups as topological invariants to spaces, this would be useful as there is considerable variety amongst groups compared with just being able to say that a metric space is compact, or not, connected, or not. Further, the connection, between the groups we will introduce and the spaces they help describe, has a *naturality* to it as we'll see continuous functions between two spaces yield algebraically nice functions between their associated groups.

Algebraic topology

Our aim now is to associate with a space a group that captures something of that space's topological essence. The great French mathematician and physicist Henri Poincaré had the idea to consider *loops*, paths in the space that begin at a point and return to the same point. We might first try appreciating Poincaré's ideas on a torus. Our first problem is that there are just too many loops; two *different* loops, such as l_1 and l_3 in Figure 39(a), both go once around the hole in the torus in the same direction, but both capture much the *same* about the shape of the torus. Secondly, Poincaré was seeking to define a group and so he needed to find ways to combine loops with some operation.

Addressing this second point, we could combine loops by going around one loop after going around the other; at least we could if the second loop began where the first loop ended. So, first, we choose a fixed point of the torus, a *base point*, and consider only loops that begin and end at the base point; then, given two such loops l_1 and l_2, their product $l_2 * l_1$ is, reading from the right, the path that starts at the base point, follows l_1 and then l_2, still ultimately returning to the base point.

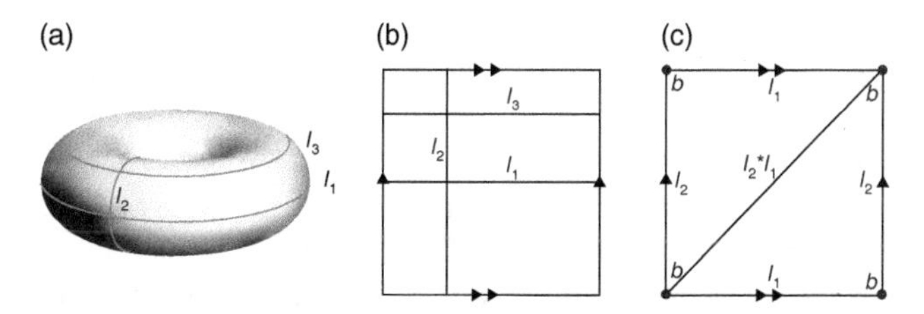

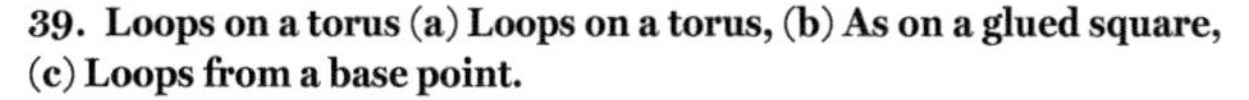
39. Loops on a torus (a) Loops on a torus, (b) As on a glued square, (c) Loops from a base point.

Would the set of loops based at a point, together with this operation $*$, make a group? What, for example, would the identity loop e be, as described in rule 2? This loop e would have to be such that $l * e = l$ for all loops l. This means that the path l followed by e would have to be the same as just the path l and so e would need to be the path that goes nowhere—e starts at the base point and immediately finishes at the base point.

And what would the inverse l^{-1} of a loop l be, as described in rule 3? The inverse l^{-1} of l must satisfy $e = l^{-1} * l$, knowing that e is the 'don't move' path at the base point. This is problematic; even if we defined l^{-1} to be the return journey of l, that is going around l in the opposite direction, e and $l^{-1} * l$ would not be equal. To travel from London to Paris and back by the same route is in no obvious way the same as just staying put in London.

Poincaré was able to resolve this issue with the same idea that he resolved the first problem, that of there being too many loops. The two loops l_1 and l_3 in Figure 39(a) seem to capture the same aspect of the topology of a torus, but they're different loops. However, either of them could be continuously deformed into the other. The technical term for this is that the two loops are **homotopic**. Any loop in the torus that is homotopic to l_1 in Figure 39(a) is a loop that goes once around the torus's hole in the same direction. But l_2, a loop which goes once through the hole, is not homotopic to either l_1 or l_3. In Figure 39(b) these loops are again depicted on a torus represented as a glued square.

Why would it now be that following a path l, and returning along that same path, is homotopic to making no journey at all? You might think of a journey that did 99 per cent of the path l and returned from there, and then one that did the same for 98 per cent, etc. Hopefully you now have the idea that you can gradually do less and less of that return trip to Paris to the point where you just stay in London—a kind of spectrum of increasingly frustrating weekend breaks!

Poincaré introduced the **fundamental group** of a space in 1895 in his seminal work *Analysis Situs* (the title being an old, now obsolete, name for topology). The elements of the fundamental group are loops based at some point in the space, but with the understanding that two loops that can be continuously deformed into one another are to be considered the same. In the case of the torus—or indeed of any path-connected space—the fundamental group doesn't depend on our choice of base point.

In Figure 39(c) a base point b is shown, and two loops l_1 and l_2, similar to before but now based at b. A loop (based at b) that goes once around the hole of the torus is homotopic to l_1 and these loops are equal in the fundamental group. A loop that goes once through the hole is homotopic to l_2. In fact it can be shown that any loop in the torus (based at b) is homotopic to a journey around l_1 some number of times followed by the journey around l_2 a certain number of times. What is being claimed here is that knowing how many times a loop wraps around the torus's hole and how many times it wraps through the hole determines the essence (up to homotopy) of a loop on the torus.

There is a subtle, implicit point here worth noting. The order in which the loops wrap around or through the hole does not matter on the torus (but does for other surfaces). This is because $l_1 * l_2 = l_2 * l_1$. This is easiest seen in Figure 39(c). l_1 is the loop running along the base (or top) of the square and l_2 is the loop running along the left (or right) side of the square. The loop $l_2 * l_1$ is then the loop starting in the bottom left corner and going right along the bottom and then up the right side, and $l_1 * l_2$ is the loop starting in the bottom left corner and going up the left side and then right across the top. We can see that $l_1 * l_2$ and $l_2 * l_1$ are homotopic—and so equal in the fundamental group—by imagining the right-then-up path being continuously dragged across the square until it becomes the up-then-right path. The square's diagonal, as in Figure 39(c), is a loop equal to either product.

Because $l_1 * l_2 = l_2 * l_1$, a loop like $l_2 * l_1 * l_1 * l_2 * l_2 * l_1 * l_1$ in the fundamental group is the same as $l_1 * l_1 * l_1 * l_1 * l_2 * l_2 * l_2$—the order that the loops are travelled does not matter, only that four l_1s and three l_2s are followed. A loop on the torus is essentially characterized by how many times it wraps around the hole of the torus (as l_1 does) and through the hole of the torus (as l_2 does). The full consequence of all this is that the fundamental group of a torus is $\mathbb{Z}^2$, the group of pairs of whole numbers, with (m, n) representing those loops going around the hole m times and through the hole n times.

It may not now be that surprising to find that the fundamental group of a circle is $\mathbb{Z}$, the group of whole numbers under addition, as a loop in the circle is essentially characterized by how many times it wraps around the circle. Let's remind ourselves of just what's being claimed: a loop that goes around the circle, starting and finishing at the same point, goes around the circle a certain number of times (counted in an anti-clockwise sense). Another, different loop may go around the loop the *same* number of times; this second journey might have digressions here and there, back-and-forth, but because the two journeys overall go around the circle the same number of times then one journey can bit-by-bit be deformed into the second journey; if the first journey gets behind/ahead of the second journey then it can become increasingly hurried/delayed to rectify that. But a loop that goes around the circle twice anti-clockwise could never be deformed into one that goes once clockwise; these two journeys are essentially different—that is, they're not homotopic. These two journeys correspond to the numbers 2 and –1 in the circle's fundamental group. Any deformation of a loop that goes twice around the circle will always be another loop that still goes twice around the circle.

Finally, to say that the fundamental group of a circle is the group of whole numbers under addition, means that loops around the circle combine in the same way integers add. In the group $\mathbb{Z}$ of whole numbers, we know that the numbers 2 and –1 add to give 1.

In the fundamental group, 2 represents any loop that goes twice anti-clockwise around the circle and −1 represents any loop that goes once clockwise around the circle. If we combine such loops using $*$ then we get a loop that goes in total once anti-clockwise around the circle and such a loop corresponds to 1 in the group of whole numbers. So not only do the whole numbers correspond to loops around the circle, but how whole numbers add corresponds to how loops combine using $*$.

In Chapter 1, we mentioned that the letter O is not *simply connected*, meaning O 'has a hole in it'. Unsurprisingly neither the circle nor torus are simply connected either. But we can now formally say what it is for a space to be **simply connected**, which means that the space is path-connected with fundamental group $\{e\}$. This means that every loop is homotopic to the constant loop e. So lines, planes, discs are simply connected; cylinders and Möbius strips aren't as a loop going once around either is not homotopic to e. In general, a loop going around a hole in a space—like an elastic band caught around a stick—cannot be transformed to a constant loop. The elastic band analogy is a useful one: the plane with a point removed is not simply connected—an elastic band can be hooked around the missing point—but 3D space minus a point *is* simply connected as an elastic band cannot be hooked by a single missing point in 3D. But 3D space missing a line *isn't* simply connected as the elastic band can become irremovably hooked on the line.

A further important feature of fundamental groups is how a continuous function between two spaces leads to a nice algebraic function between their fundamental groups. Say f is a continuous function between two spaces X and Y. If l is a loop in a space X based at a point b, then $f(l)$ is a loop in the space Y based at the point $f(b)$. Take a moment to appreciate this: f takes points of X to points of Y and l is a journey through such points of X beginning and ending in b. So $f(l)$ is a journey of points in Y beginning and ending at $f(b)$.

As an example, consider when f is the function which wraps the circle twice on to itself as shown in Figure 40(a). A point making angle θ with the horizontal is sent to a point making angle 2θ with the horizontal axis—so the arc from a to p would map to the arc from a to $f(p)$, and the upper semicircle from a to b would map to the whole circle. And if we had a loop l in the circle that goes once (anti-clockwise) around the circle then $f(l)$ goes twice around the circle; if l goes twice around the circle then $f(l)$ goes four times around the circle. Loops that go n times around the circle are what n represents in the fundamental group of the circle, and these loops are sent to loops that go $2n$ times around the circle, which are what $2n$ represents in the fundamental group. The function f wraps the circle twice onto itself and naturally gives a corresponding function f for loops, with a loop l that wraps n times around the circle being sent to a loop $f(l)$ that wraps $2n$ times around the circle.

Say l_1 and l_2 are loops that wrap around the circle n_1 and n_2 times, so that $l_2 * l_1$ is a loop that wraps around the circle $n_2 + n_1$ times. So $f(l_1), f(l_2), f(l_2 * l_1)$ are loops that wrap around the circle $2n_1$, $2n_2$, and $2(n_2 + n_1)$ times; similarly $f(l_2)*f(l_1)$ wraps around the circle $2n_2 + 2n_1$ times. As

$$2(n_2 + n_1) = 2n_2 + 2n_1$$

then both $f(l_2 * l_1)$ and $f(l_2)*f(l_1)$ wrap around the circle an equal number of times, meaning they are homotopic—that is

$$f(l_2 * l_1) = f(l_2) * f(l_1) \qquad \text{(H)}$$

as elements of the fundamental group.

More generally a continuous function f between two spaces X and Y naturally gives a corresponding function f from the fundamental group of X to the fundamental group of Y, as a loop l based at x is

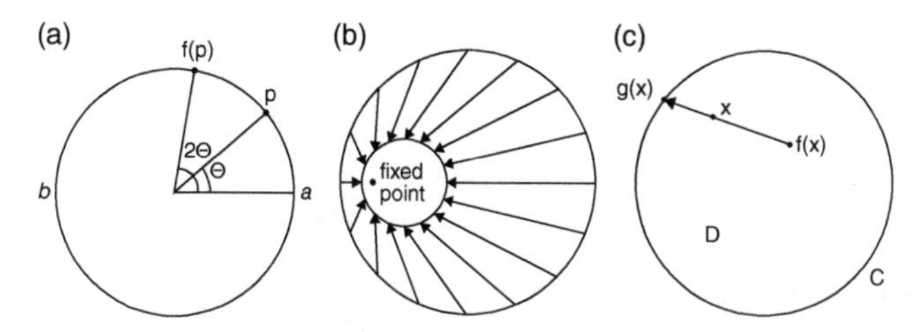

40. Functions on the disc (a) *f* wrapping the circle twice around, (b) A function *f* from *D* to *D*, (c) Defining the function *g*.

sent to a loop $f(l)$ based at $f(x)$ and this function satisfies (H) in general. Any function between groups that satisfies (H) is called a **homomorphism**. For groups, homomorphisms are the natural functions to study, being the functions that respect group operations.

We now use these ideas to prove **Brouwer's fixed point theorem**, first proved by L. E. J. Brouwer in 1910, which states:

> Let D be a closed disc and f be a continuous function with inputs and outputs in D.
>
> Then f has a fixed point: there is some x in D such that $f(x) = x$.

The theorem applies to a closed disc D, a disc including its circumference. (It's not hard to construct a function from an open disc to itself with no fixed points.) There are many continuous functions from D to D. Such a function is rotation of D about its centre, and in this case the only fixed point is the centre. In Figure 40(b), the disc has been shrunk and moved somewhat to the left. Again, the theorem states that there is a fixed point (which is unique again in this case). It's not hard to visualize lots of different ways that D might be transformed into D and, in all cases, there must be at least one fixed point.

The theorem's proof is not easy, perhaps involving the most conceptual ideas in this text. First the proof is *by contradiction* which means that we consider the possibility of there being a function f with no fixed points and argue from that assumption to a position that is logically impossible. If a function f has no fixed points, then $f(x) \neq x$ for all x in D. Because of this we can define the function g, from D to its circumference C as follows (Figure 40(c)): as $f(x)$ and x are different points, then we can draw a line starting from $f(x)$ and going through x which eventually meets the circumference C at the point $g(x)$. This way any point x in the disc D has been sent to a point $g(x)$ that lies on its circumference C. Note that if x is itself on the circumference C, then $g(x) = x$.

So, there are now several functions we can consider:

- i, inclusion, which has inputs in C and outputs in D;
- g which has inputs in D and outputs in C;
- the composition $g \circ i$, which has inputs in C and outputs in C.

Here *inclusion* means the function which takes a point of C, the circumference, as an input, and returns the same point as the output, now considered as a point of the disc D. The *composition* $g \circ i$ means the function which performs inclusion i first and then performs g second. Earlier we noted g fixes points on the circumference C and so, importantly, the function $g \circ i$ fixes all points of C.

We arrive at our contradiction by considering the corresponding functions on the fundamental groups of C and D. Recall that the fundamental group of C, a circle, is the group of whole numbers, with the number n representing those loops that wrap n times anti-clockwise around the circle; the fundamental group of the disc D, which is simply connected, is $\{e\}$. The corresponding functions for the fundamental groups satisfy the following:

- i has inputs in the whole numbers and an output of e, whatever the input, so that this function is constant;
- g has a single input e and so some single output in the whole numbers;
- the composition $g \circ i$, which has inputs and outputs in the whole numbers.

Remember that the composition $g \circ i$ fixes all points of C and so any loop in C is sent to the same loop by the corresponding function $g \circ i$ on loops. So, in terms of the fundamental group of C, the function $g \circ i$ sends each whole number n to itself. Instead we can consider the separate effects of doing the corresponding function i first and then the corresponding function g. As noted, i sends all the whole numbers to e, which g subsequently sends to $g(e)$. Note that this number $g(e)$ is a single whole number not depending on n.

Looked at one way, the function $g \circ i$ on loops sends all whole numbers n to themselves, but considering the functions' effects separately, every n is sent to $g(e)$; whatever this value $g(e)$ is, it is a single value and so $g \circ i$ is a constant function. This is the required contradiction—the function $g \circ i$ cannot both send each n to n and also send each n to a single value $g(e)$. So a function f, without fixed points, cannot exist or else we'd be able to define the function g and arrive at a contradiction.

The ideas of this proof are subtle, but the crux of it, and what makes such methods powerful, is the following: fundamental groups (and other similar algebraic invariants) are able to capture something of the topological essence of a space; continuous functions between spaces lead to algebraically nice functions (homomorphisms, that have the earlier property H) between the spaces' fundamental groups; if it can be shown no such homomorphism exists, then no such continuous function existed.

There are further algebraic topological invariants capturing the essence of a space in higher dimensions. The plane with a point missing is not simply connected as a loop that goes around the missing point cannot be 'unhooked'. 3D space missing a point *is* simply connected though—an elastic band (a metaphorical circle) in 3D cannot be hooked by a single missing point, but we can imagine a balloon (a metaphorical sphere) being hooked onto the missing point. So there is something topological occurring in punctured 3D space, but in a higher dimension than the fundamental group, with its loops, can measure.

Some such information can be captured by **Betti numbers,** named by Poincaré after Enrico Betti who first studied these. These Betti numbers are topological invariants and an n-dimensional space has Betti numbers $b_0, b_1, b_2 \ldots b_n$ with the ith Betti number b_i capturing something of the essence of the space's topology in the ith dimension. $\mathbb{T}^{\#g}$, the torus with g holes, has Betti numbers

$$b_0 = 1, \qquad b_1 = 2g, \qquad b_2 = 1,$$

where b_1 captures there being $2g$ loops in the torus, one around each of the g holes of the torus, and one in and through each hole. And the Betti numbers of $\mathbb{P}^{\#k}$ are

$$b_0 = 1, \qquad b_1 = k - 1, \qquad b_2 = 0.$$

The Euler number of a space is defined in terms of its Betti numbers as the alternating sum

$$b_0 - b_1 + b_2 - b_3 + b_4 - \dots,$$

so that the Euler number of $\mathbb{T}^{\#g}$ equals $1 - 2g + 1 = 2 - 2g$ and of $\mathbb{P}^{\#k}$ equals $1 - (k - 1) + 0 = 2 - k$. That these Betti numbers are topological invariants then means that the Euler number is a topological invariant.

Poincaré is often considered as the last universalist in mathematics, someone making research contributions across mathematics, and he became interested in topology via various routes. In 1889 he won a mathematical competition funded by Oscar II, King of Sweden and Norway, for contributions to understanding how a multiple body system (like the planets in the solar system under gravity) evolves with time. So, for Poincaré, topology was a means for qualitatively analysing such systems. In pure mathematics, he was interested in the topology of spaces in their own right, essentially studying manifolds though still lacking at the time a wholly rigorous definition for such spaces.

It was also Poincaré who introduced **combinatorial topology** to help calculate algebraic topological invariants. As I mentioned in Chapter 1, there are many complicated examples of Jordan curves but, ultimately, all of them are homeomorphic to a triangle; likewise the sphere's topology is no different from a cube, nor a torus's from the polyhedron in Figure 11(a). By using polygons, polyhedra, and higher-dimensional equivalents, quite general spaces can be approximated by simpler polyhedra that can be finitely described but still capture all of their topological essence. But, again, Poincaré's ideas were far-sighted but not entirely rigorously realized.

For around the next forty years, topologists would take forward and make rigorous Poincaré's vision, making algebraic topology a major theme of mathematics and fully appreciating the range and power of his ideas, with James Alexander—more of him in Chapter 6—perhaps being most prominent in that role. Arguably the last piece of Poincaré's legacy came with the proof in 2003 of the **Poincaré conjecture** by Grigory Perelman. The conjecture states that every simply connected, compact three-dimensional manifold is homeomorphic to the three-dimensional sphere (this does not mean a solid ball in 3D, but rather a 3D spherical shell that sits naturally in 4D). His proof, based on a strategy developed by Richard Hamilton, used ideas of differential geometry far beyond even Poincaré's imagination at the time. However, its solution is more evidence for how the great problems of mathematics have led to progress in highly novel directions.

Chapter 6
Unknot or knot to be?

Describing knots

A **knot** is a smooth, simple, closed curve in 3D space. Being simple and closed means the curve does not intersect itself except that its end returns to its start. Basically a knot is a loop in 3D space and by requiring smoothness we exclude some nasty, so-called *wild*, knots from our study. As I noted in Chapter 1, all knots are topologically the same as a circle; what makes a circle knotted—or not—is how that circle has been placed into 3D space.

As knots—in and of themselves—are just circles, then we need a new notion to say when two knots are the same, as situated in 3D space. That notion needs to capture the idea that one knot, and the space around it, can be continuously deformed into the other, and the space around it; this is called an *ambient isotopy*.

So, two knots K_1 and K_2 are to be considered equivalent if we can start with knot K_1 and over a period of time continuously deform 3D space so that K_1 becomes K_2. If we set the time taken to deform the knots as a unit interval $0 \leqslant t \leqslant 1$, then by some time t in the middle of that interval we will have deformed 3D space by some homeomorphism h_t. An ambient isotopy is then a continuous family of such h_t, where at time $t = 0$ we have yet to start deforming so that $h_0(x) = x$ and by the time $t = 1$ the first knot has been deformed into the second, that is $h_1(K_1) = K_2$.

The central problem of knot theory is then a classification theorem: when are two knots *equivalent*—there is an ambient isotopy between them—or how do we show that no such isotopy exists? This problem was first identified by Maxwell in 1868, though his work went unpublished at the time.

A more basic problem of knot theory is first a means of describing knots. We can imagine laying any knot flat on a table top. Not literally flat—the essence of the knot will be in how at certain crossings one part of the knot goes over or under another bit of the knot (Figure 41); we can also be careful to separate out those crossings so that no more than two parts of the knot cross at the same place. The *minimal* number of crossings of a knot is a *knot invariant*, so that equivalent knots have the same minimal number of crossings. In Figure 41(a) appears a version of the **unknot**; by an unknot we mean a loop that is not actually knotted, and so is isotopic to a circle. Figure 41(a) has two crossings but the *same* unknot can be represented as a circle with zero crossings. The first significant effort at describing and classifying knots by means of over-and-under crossings was by Peter Guthrie Tait, between 1877 and 1885, who classified all knots with ten or fewer crossings.

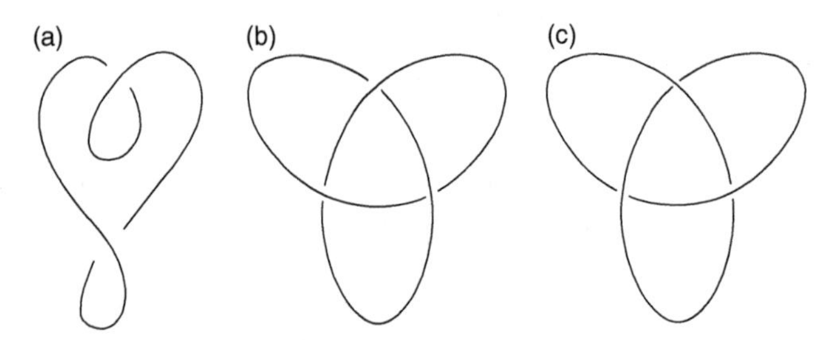

41. The unknot and trefoils (a) An unknot, (b) Left-handed Trefoil, (c) Right-handed Trefoil.

Knots only exist in 3D; any loop in 4D can be unknotted. Beginning from the rightmost point of the unknot in Figure 41(a), and moving anti-clockwise around the loop, we see that the two crossings both go over the other part of the knot. More generally any loop that we could lay flat on the table, at each crossing laying the loop over a previous part of the loop, would lead to an unknot, as we could lift the loop off the table one crossing at a time to make a circle. Imagine a genuine knot, with various under- and over-crossings, but with us now permitted to move in 4D. Much as we earlier used time as the fourth dimension to avoid the Klein bottle intersecting itself, we could use this extra dimension to make any under-crossing into an over-crossing; we could take the under-part of the knot smoothly into the future, raise it above the over-part of the knot which only exists in the present, and then smoothly return the knot from the future to the present, now as an over-part. In this way all knots could be unknotted in 4D.

The **trefoil** is the simplest genuine knot and has three crossings. (You might want to sketch for yourself loops with one or two crossings and convince yourself these aren't proper knots.) The trefoil though is complicated enough to exhibit an important feature, that of **chirality**. The trefoil knots in Figures 41(b) and 41(c) are mirror images of one another and *aren't* isotopic to one another. The trefoil is also an example of an **alternating knot**, a knot where the crossings alternate between over- and under-crossings as we travel around the knot. Alternating knots provide a somewhat simpler class of knots which are well understood compared with knots generally.

Reidemeister moves

The unknot in Figure 41(a) can be quickly deformed into a circle by untwisting it at the top, and doing similar to the twist at the bottom. Such a move—a *twist* or *untwist*—is the first of three **Reidemeister moves** that between them can be used to show any equivalent knots are indeed equivalent. These moves are named

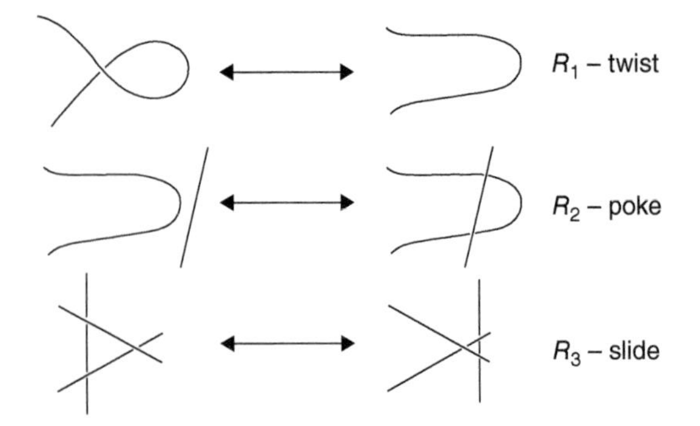

42. The three Reidemeister moves.

after Kurt Reidemeister, who proved this result in 1927, though the same result had been independently proved by Alexander and Briggs in 1926, and in fact the moves were known much earlier to Maxwell.

The first move then, R_1, is a twist or untwist of a small part of the knot (Figure 42). The second move R_2 takes a small part of the knot and *pokes* it under another part of the knot or undoes this. Finally, the third move R_3 is to *slide* a single crossing over a small part of the knot. Hopefully none of these moves seems controversial and are evidently permitted manipulations of the knot. The significant result is that, given any two equivalent knots, repeated use of these three moves alone is sufficient to show that the two knots are indeed equivalent.

Unfortunately, many theorems in mathematics demonstrate the *existence* of a solution, without being *constructive*—that is, providing a means to find that solution—or, even better, providing an *efficient* constructive means to finding a solution. In 1961 Wolfgang Haken showed that the *unknotting problem* is *decidable*, in the sense that there is a general algorithm that can be implemented to decide whether a knot is the unknot. Others would later extend this result to show the *equivalence problem* is decidable, that there is an algorithm for checking whether two knots are equivalent. But it is still an open problem as to whether there is an efficient algorithm for this problem.

Prime knots and adding knots

In Figure 43 appear the *granny knot* and the *reef knot*. Notice how the granny knot is reminiscent of two merged trefoil knots—in fact, it is the *connected sum* of two trefoil knots with the same handedness, whilst the reef knot is the connected sum of a trefoil with its mirror image. As in Chapter 2 with surfaces, we can create the connected sum $K_1\#K_2$ of two knots K_1 and K_2 by removing a small arc of each knot and gluing opposite loose ends together.

If K_1 and K_2 have n_1 and n_2 minimal crossings, then $K_1\#K_2$ can clearly be drawn with n_1+n_2 crossings. However, it is a currently unsolved problem as to whether the *minimal* number of crossings for $K_1\#K_2$ equals n_1+n_2. This has been shown to be true for alternating knots, and is suspected to be true in general, but remains unresolved.

As our central problem is to classify knots then it makes sense to focus on those knots that are not connected sums of simpler knots. Such knots are called **prime knots**. Not counting mirror

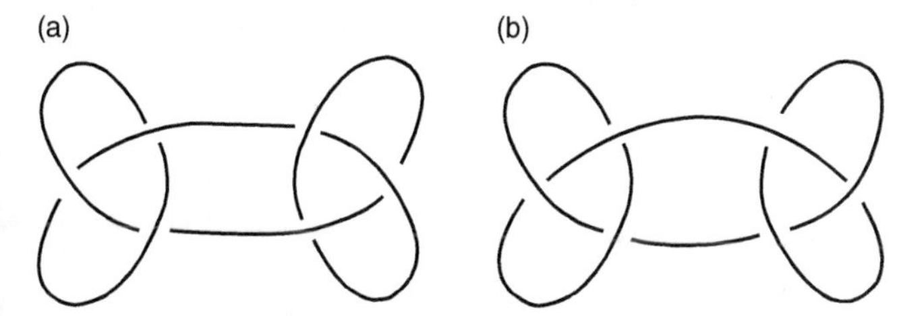

43. The granny and reef knots (a) A granny knot, (b) A reef (or square) knot.

images, then the numbers of prime knots with crossing number of 12 or less are given in Table 6.

Note how this number of different prime knots spirals enormously as the minimal crossing number increases. It is an open and active area of research seeking to prove *asymptotic* estimates for how quickly the number of prime knots grows as the crossing number becomes large. Figure 44 depicts these prime knots up to a minimal crossing number of seven, all of which are alternating; the simplest non-alternating prime knots have a minimal crossing number of eight. In 1998 asymptotic estimates were demonstrated for prime, alternating knots; the same result also showed that

Table 6. Number of prime knots sorted by minimal crossing number

Minimal crossing number	3	4	5	6	7	8	9	10	11	12
Number of prime knots	1	1	2	3	7	21	49	165	552	2176

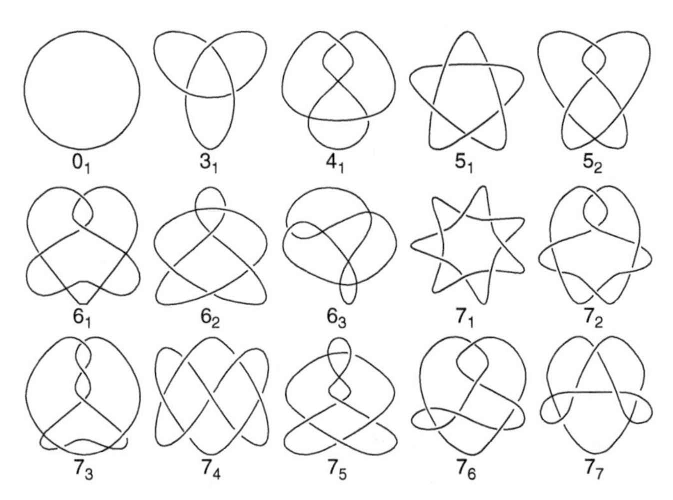

44. Prime knots with seven or fewer crossings.

alternating knots become increasingly rare, as a fraction of all knots, as the minimal crossing number increases.

The knot group

Reidemeister moves provide a difficult-to-implement means for deciding whether two knots are equivalent or not, and so topologists seek *knot invariants* that are easy to calculate but, ideally, separate out inequivalent knots. All knots are, in themselves, just circles—it's how that circle has been situated in 3D space that makes it a particular knot. So we might focus instead on the *complement* of the knot, the remainder of 3D space that isn't the knot, to better understand knots. We might consider the fundamental group of the complement, and this is known as the **knot group**.

Recall that the elements of a fundamental group are loops based at a point, with loops understood to be the same if one can be deformed into the other. So, starting from and finishing at a point outside a knot, these different loops might weave in and out of a knot helping to capture something of the essence of the knot. And this is indeed the case, but unfortunately knot groups remain very complicated objects. The knot group of the unknot (a circle) is just the group of whole numbers, with a loop being entirely characterized by how many times it wraps in and through the unknot. But even the knot group of the trefoil is a difficult group to describe.

Wilhelm Wirtinger, around 1905, found a way to describe such knot groups in general. In Figure 45 is drawn an oriented trefoil with the three unbroken arcs between the crossings labelled as a_1, a_2, a_3 and the five regions the trefoil splits the plane into denoted $R_1 \ldots R_5$. Imagine the base point b of the knot group being outside the trefoil, and loops beginning at the base point, weaving in and out of the knot, and returning back to the base point. We will write l_1 for a loop that goes down through R_3 and returns back out of R_5, or equally down through R_1 and returns back out of R_4. This is a loop from the base point b that 'hooks' a_1 going in on the left of

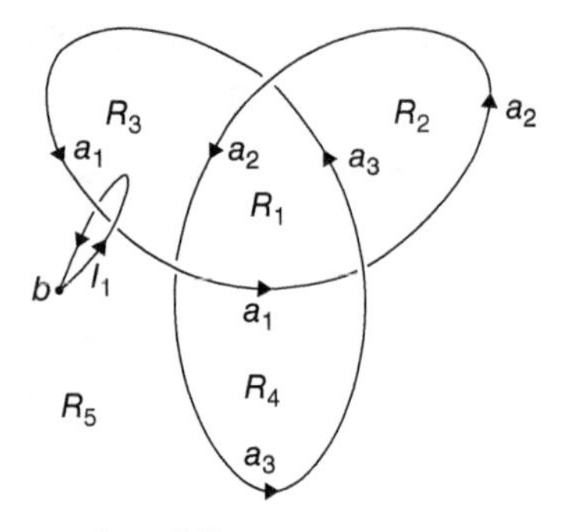

45. The knot group of a trefoil.

a_1 and coming back out on the right of a_1. We denote by l_2 and l_3 similar loops that hook a_2 and a_3 in a left-to-right manner. Their inverses l_1^{-1}, l_2^{-1}, l_3^{-1} are loops that hook a_1, a_2 and a_3 in the reverse right-to-left direction. Any loop beginning and ending in the base point b can be written as a string using the symbols l_1, l_2, l_3, l_1^{-1}, l_2^{-1}, l_3^{-1} such as

$$l_3 l_2 l_1^{-1} l_2 l_1 l_1^{-1} l_3 l_2^{-1} l_3 l_2 l_1,$$

a recipe for how the loop weaves in and out of the knot, with careful attention as to whether the loop went in-left and out-right or vice versa. But already we see that this can't be the whole description of the knot group: in the middle of the above 'recipe' is the expression $l_1 l_1^{-1}$ which cancel out one another (remembering back to Chapter 5 and the diminishing London–Paris return trip). To do a loop and then do it in reverse is essentially the same as not moving. So, we should omit all occurrences of expressions like $l_1 l_1^{-1}$, $l_2^{-1} l_2$, etc.

However, there are other 'relations' between these loops l_1, l_2, l_3 and there is one such relation for each crossing of the knot. Consider a loop from b that hooks the knot at the crossing between R_1, R_3, R_4, R_5; this is a loop that goes down through R_1 and back out through R_5. This can be achieved as $l_3 l_1$, which means doing l_1 first and l_3 second; so we go in through R_1 and return out of R_4 and then back in through R_4 and out of R_5. Or we could manage the same by doing $l_1 l_2$ which means doing l_2 first and l_1 second; this takes us in through R_1 and returns out of R_3 and then down through R_3 and out of R_5. Either of these loops hooks the knot at the crossing point and so it's the case that $l_3 l_1 = l_1 l_2$.

Now if there were n crossings in a knot, then we'd have strings involving $l_1 \ldots l_n$ and $l_1^{-1} \ldots l_{n-1}$ in the knot group. Wirtinger's theorem showed that the knot group consists of all such strings, with two such strings describing the same loop if one string can be turned into the other by cancelling terms like $l_1 \, l_1^{-1}$ or using rules, like $l_3 l_1 = l_1 l_2$, with one such rule coming from each crossing.

All this is impressive, given its generality, but is not very tractable given our aim is to describe simple invariants to separate out inequivalent knots. It's also frustrating to find that inequivalent knots, such as the reef knot and the granny knot, can have the same knot groups, as do left- and right-handed trefoils. However, it was shown in the late 1980s that the knot group determines a *prime* knot up to mirror images: two prime knots with the same knot group are isotopic knots, or each knot is isotopic to the mirror image of the other.

Alexander and Jones polynomials

In 1928 James Alexander introduced what is now known as the **Alexander polynomial** and which, for a knot K, is usually denoted $\Delta_K(x)$. Here x is the variable of the polynomial. The Alexander polynomial is a knot invariant and much simpler than the knot group, though it ultimately conveys less information about a knot—knots with the same knot group have equal Alexander polynomials.

The Alexander polynomial for the unknot is the constant polynomial 1 and the Alexander polynomial of the trefoil T equals

$$\Delta_T(x) = x - 1 + x^{-1}.$$

So technically the Alexander polynomial is a polynomial in the variables x and x^{-1}. In his 1928 paper Alexander described an algorithm for calculating his polynomial from an over- and under-crossings description of the knot and that calculation, for the trefoil, is done in the Appendix. The algorithm is a little technical but ultimately uses mathematics taught in schools and colleges.

The Alexander polynomial also deals well with composite knots, having the nice algebraic property

$$\Delta_{K\#L}(x) = \Delta_K(x) \times \Delta_L(x)$$

for two knots K and L, with $K\#L$ denoting their connected sum. So, the reef and granny knots both have Alexander polynomial $(x - 1 + x^{-1})^2$.

The Alexander polynomial cannot distinguish between mirror images, but does distinguish between prime knots of up to eight crossings. Surprisingly there are non-trivial knots, including one with just eleven crossings that has a constant Alexander polynomial equal to 1, so the Alexander polynomial cannot distinguish the unknot from all other knots.

It was some considerable time later, 1984, when a second polynomial invariant, the **Jones polynomial**, was discovered by Vaughan Jones, for which he would win the Fields Medal in 1990. I won't define the Jones polynomial algorithmically here, but rather describe properties which characterize it uniquely.

The Jones polynomial of a knot L is denoted $V_L(x)$ and is actually a polynomial in the variables $\sqrt{x}$ and $1/\sqrt{x}$. The Jones polynomial of the unknot is 1. The following *skein relation* then

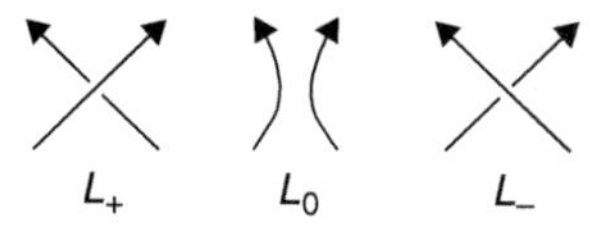

46. Links involved in the skein relation.

characterizes the Jones polynomial. (The word 'skein' means a quantity of thread or yarn.)

We consider three knots L_+, L_0, and L_- that differ only at one crossing. The knots L_+ and L_- differ in which part of the knot makes the over-crossing and the knot L_0 has no crossing at all at this point (Figure 46).

The **skein relation** then states that

$$\frac{1}{x}V_{L_+}(x) - xV_{L_-}(x) = \left(\sqrt{x} - \frac{1}{\sqrt{x}}\right)V_{L_0}(x).$$

As mentioned before, if we made all the crossings of a knot over-crossings (or all under-crossings) then we make the unknot. If in a knot an over-crossing L_+ makes that knot more 'knotted' than an under-crossing L_- then the skein relation describes the Jones polynomial of the more complicated knot L_+ in terms of the less complicated knot L_- and a knot L_0 with one fewer crossing.

We will shortly use the skein relation to calculate the Jones polynomial of the trefoil, but an important point has so far been ignored. Even if L_+ and L_- are knots, L_0 need not be. As we will see in the example of the trefoil, eliminating a crossing might disconnect a knot and create what is called a *link*, which is just a collection of knots. The skein relation, and Jones polynomial generally, should be seen as relating to such links.

The Jones polynomial can sometimes differentiate between a knot K and its mirror image K^* as the identity $V_{K^*}(x) = V_K(1/x)$ holds; so, we need to specify the trefoil being considered as right-handed.

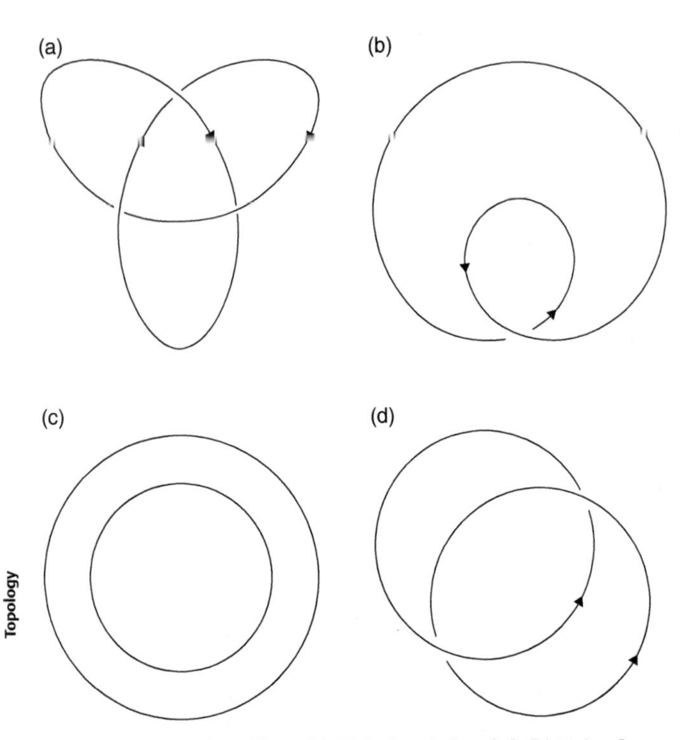

47. Simple examples of links (a) Right-handed trefoil, (b) Twisted unknot, (c) Unlinked circles, (d) Linked circles.

In our calculation we'll need to consider various knots and links (Figure 47).

We will take L_+ as the trefoil and focus on the bottom-right of the three crossings. L_- is then the unknot, perhaps easiest seen by unpoking the bottom part of the knot with a Reidemeister move. And L_0 has become disconnected making two linked circles as in Figure 47(d). Remembering that $V_{\text{unknot}}(x) = 1$ the skein relation states

$$\frac{1}{x}V_{\text{trefoil}}(x) - x = \left(\sqrt{x} - \frac{1}{\sqrt{x}}\right)V_{\text{linked}}(x),$$

where $V_{\text{linked}}(x)$ is the Jones polynomial of two linked circles. Using the skein relation a second time, we take L_+ as the linked circles, and consider the rightmost crossing. Then L_- is two unlinked circles (Figure 47(c)) and L_0 is the unknot. This time the skein relation states

$$\frac{1}{x}V_{\text{linked}}(x) - xV_{\text{unlinked}}(x) = \sqrt{x} - \frac{1}{\sqrt{x}}.$$

Finally, we apply the skein relation to the twisted unknot (Figure 47(b)) to find out $V_{\text{unlinked}}(x)$. If we take L_+ as the twisted unknot then L_- is another twisted version of the unknot and L_0 is two unlinked circles. The skein relation then states

$$\frac{1}{x} - x = \left(\sqrt{x} - \frac{1}{\sqrt{x}}\right)V_{\text{unlinked}}(x)$$

which rearranges to give

$$V_{\text{unlinked}}(x) = -\sqrt{x} - \frac{1}{\sqrt{x}}.$$

Substituting this into earlier equations gives

$$V_{\text{linked}}(x) = -\sqrt{x} - x^2\sqrt{x}$$

and then with a little rearranging we find

$$V_{\text{trefoil}}(x) = x^2 + x\left(\sqrt{x} - \frac{1}{\sqrt{x}}\right)\left(-\sqrt{x} - x^2\sqrt{x}\right) = x + x^3 - x^4.$$

So $V_{\text{trefoil*}}(x) = V_{\text{trefoil}}(1/x) = 1/x + 1/x^3 - 1/x^4$ for the trefoil's mirror image, and the Jones polynomial can differentiate between the left- and right-handed trefoils. The Jones polynomial distinguishes between prime knots with up to nine crossings, but not beyond. Unlike the Alexander polynomial it remains an unsolved problem as to whether there exists a genuine knot with a Jones polynomial equal to 1.

Epilogue

In 1911 the French mathematician Jacques Hadamard wrote that

> Analysis situs... constitutes a revenge of geometry on analysis.

(Recall that 'analysis situs' is an old name for topology.) Certainly geometric topology and the visualization of Riemann, Klein, Möbius, and Poincaré were in marked contrast to the analysis of Weierstrass. Riemann's and Poincaré's work would have massive influence on the development of mathematics. Topology would go on to become one of the central themes of mathematics.

And from the earliest history of topology, connections with physics would be apparent. In the 19th century Gauss and Maxwell would both recognize such in electromagnetism—for example, in the study of the work done by a magnetic pole moving in the presence of a wire carrying current, with Gauss's answer being in terms of a linking number for the pole's path and the wire. Both Gauss and Maxwell would bemoan the lack of progress with the study of topology or 'geometry of position' as Maxwell referred to it at the time.

The 20th century would develop a yet richer connection between topology and physics. In 1965, Roger Penrose would use topological ideas to demonstrate how the gravitational collapse of a massive star would lead to a space-time singularity occurring, such as a black hole. The use of topology meant that Penrose was able to impose qualitative assumptions about the mass distribution, compared with earlier assumptions about the symmetric distribution of matter that had been considered physically questionable.

The interaction between physics and topology would also not be one way. A problem that was ostensibly in physics would become of interest to pure mathematicians if it could be rephrased into mathematical language involving mathematical objects. This was particularly the case with Yang–Mills theory, a physical theory seeking to provide a unified description for electromagnetism and the weak force. In 1983 Simon Donaldson would use ideas from Yang–Mills theory to prove astonishing results about the topology of four-dimensional manifolds.

Topology remains a large, active research area in mathematics. Unsurprisingly its character has changed over the last century—there is considerably less current interest in general topology, but whole new areas have emerged, such as *topological data analysis* to help analyse big data sets. The interfaces of topology with other areas, including physics, have remained rich and numerous, and it can be hard telling where topology stops and geometry or algebra or analysis or physics begin. Often that richness comes from studying structures that have interconnected flavours of algebra, geometry, and topology, but sometimes a result, seemingly of an entirely algebraic nature say, can be proved by purely topological means. In the words of Poincaré

> Mathematics is the art of giving the same name to different things

and the rise of topology has certainly helped demonstrate the interconnectedness of mathematics.

Appendix: Calculating an Alexander polynomial

Calculating the Alexander polynomial of a knot is a little technical but involves only mathematics that might be met at school or college, a knowledge of *matrices* and *determinants*. The method below was given by Alexander in 1928.

An oriented knot K, with n crossings $c_1, c_2 \ldots c_n$, divides the plane into n+2 regions, $R_1, R_2 \ldots R_{n+2}$, including the outside of the knot. Figure 48 shows a trefoil labelled in this manner.

From such a diagram we create a matrix with n rows (corresponding to the n crossings) and n+2 columns (corresponding to the n+2 regions). The entries of this matrix are then filled according to the following rules:

- If the region is not adjacent to the crossing, the entry is 0.
- If the region is on the left before under-crossing, the entry is $-x$.
- If the region is on the right before under-crossing, the entry is 1.
- If the region is on the left after under-crossing, the entry is x.
- If the region is on the right after under-crossing, the entry is −1.

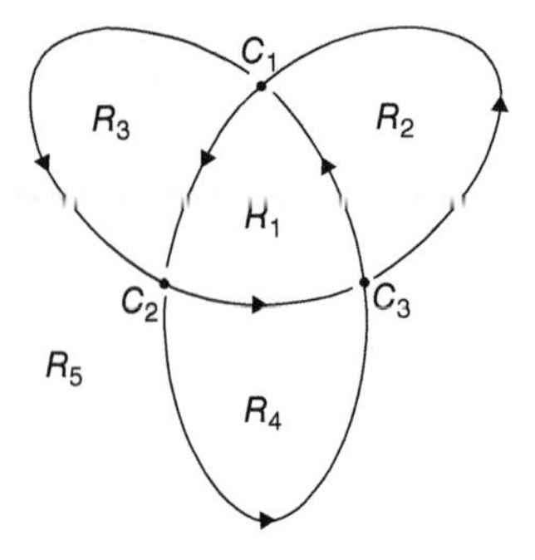

48. Calculating the Alexander polynomial of a trefoil.

This matrix for the trefoil in Figure 48 is

	R_1	R_2	R_3	R_4	R_5
c_1	$-x$	1	x	0	-1
c_2	$-x$	0	1	x	-1
c_3	$-x$	x	0	1	-1

To explain the first row, R_4 is not adjacent to c_1 and whilst travelling towards c_1 *as an under-crossing*, the region R_1 is on the left and R_2 is on the right and having passed through c_1 the region R_3 is on the left and R_5 is on the right.

We now remove any two columns corresponding to *adjacent* regions, say R_1 and R_2 in this case, so that we have a square n by n matrix, and we take the determinant of this matrix. For the trefoil this gives

$$\det\begin{pmatrix} x & 0 & -1 \\ 1 & x & -1 \\ 0 & 1 & -1 \end{pmatrix} = -x^2 + x - 1$$

The answer at this point depends somewhat on which columns we removed. If we divide by a power of x so that the highest power of x is the same as the highest power of x^{-1} (in our example we need to divide by x) and multiply by ± 1 so that the polynomial takes value 1 at $x = 1$, then we have calculated the Alexander polynomial. In the case of the trefoil knot, that equals $x - 1 + x^{-1}$.

Historical timeline

1639	Descartes's work on angular defect, equivalent to Euler's formula
1750	Euler discovers his formula $V - E + F = 2$
1817	Bolzano defines continuity, proves the Bolzano–Weierstrass and intermediate value theorems
1836	The word 'topology' is coined by Johann Listing
1851	Schläfli generalizes Euler's formula to higher dimensional polyhedra
1857	Riemann's paper *Theory of Abelian Functions* makes Riemann surfaces more widely known
1858	Möbius and Listing independently discover the Möbius strip
1861	Weierstrass lectures on the boundedness theorem
1861	Möbius gives a first sketch proof of the classification theorem for closed surfaces
1868	Maxwell first states the knot classification problem
1871	Betti publishes *On Spaces of any Number of Dimensions*
1874	Klein shows orientable closed surfaces are homeomorphic if and only if they have equal genus
1876	Klein gives his definition of one-sidedness
1881	Poincaré's theorem, later generalized by Hopf in 1926
1882	Klein first describes the Klein bottle
1887	Jordan curve theorem appears in his *Cours d'analyse*

1895	Borel states a first version of the Heine–Borel theorem
1895	Poincaré defines the fundamental group in his *Analysis Situs*
1906	Fréchet defines metric spaces in his doctoral thesis
1907	Max Dehn and Poul Heegaard give the first rigorous proof of the classification theorem
1910	Brouwer proves his fixed point theorem
1914	Hausdorff defines topological spaces in the seminal *Grundzüge der Mengenlehre*
1919	Hausdorff dimension introduced, which can take fractional values
1925	Morse publishes his paper *Relation between the Critical Points...*
1927	Reidemeister moves introduced
1928	Alexander introduces his polynomial knot invariant
1930	Kuratowski's theorem on planar graphs
1932	Čech defines higher homotopy groups, generalizing the fundamental group
1936	Whitney defines manifolds and proves his embedding theorem
1950	Hamming distance introduced in the paper *Error Detecting and Error Correcting Codes*
1952	Moise shows that every three-dimensional manifold is uniquely smoothable
1961	Haken shows that the unknotting problem is decidable
1982	Thurston wins Fields Medal for work on three-dimensional manifolds
1984	Jones polynomial introduced
1986	Donaldson and Freedman win Fields Medals for their work on four-dimensional manifolds
2003	Perelman proves the Poincaré conjecture

Further reading

Colin Adams, *Knot Book* (2004) American Mathematical Society
Jeremy Gray, *Henri Poincaré: A Scientific Biography* (2012) Princeton
Stephen Huggett and David Jordan, *A Topological Aperitif* (2009) Springer
Matt Parker, *Things to Make and Do in the Fourth Dimension* (2014) Penguin
V. V. Prasolov, *Intuitive Topology* (1994) American Mathematical Society
David Richeson, *Euler's Gem: The Polyhedron Formula & the Birth of Topology* (2012) Princeton

Online references

3Blue1Brown, *Who Cares about Topology?*
<http://www.youtube.com/watch?v=AmgkSdhK4K8&t=236s>
The Geometry Junkyard, *20 Proofs of Euler's Formula*
<http://www.ics.uci.edu/~eppstein/junkyard/euler/all.html>

Index

Note: For the benefit of digital users, indexed terms that span two pages (e.g., 52–53) may, on occasion, appear on only one of those pages.

F

G

H

I

J

K

L

M

O

P

R

S

T

U

W

ECONOMICS

A Very Short Introduction

Partha Dasgupta

Economics has the capacity to offer us deep insights into some of the most formidable problems of life, and offer solutions to them too. Combining a global approach with examples from everyday life, Partha Dasgupta describes the lives of two children who live very different lives in different parts of the world: in the Mid-West USA and in Ethiopia. He compares the obstacles facing them, and the processes that shape their lives, their families, and their futures. He shows how economics uncovers these processes, finds explanations for them, and how it forms policies and solutions.

> 'An excellent introduction ... presents mathematical and statistical findings in straightforward prose.'
>
> Financial Times

www.oup.com/vsi

INFORMATION

A Very Short Introduction

Luciano Floridi

Luciano Floridi, a philosopher of information, cuts across many subjects, from a brief look at the mathematical roots of information - its definition and measurement in 'bits'- to its role in genetics (we are information), and its social meaning and value. He ends by considering the ethics of information, including issues of ownership, privacy, and accessibility; copyright and open source. For those unfamiliar with its precise meaning and wide applicability as a philosophical concept, 'information' may seem a bland or mundane topic. Those who have studied some science or philosophy or sociology will already be aware of its centrality and richness. But for all readers, whether from the humanities or sciences, Floridi gives a fascinating and inspirational introduction to this most fundamental of ideas.

'Splendidly pellucid.'

Steven Poole, The Guardian

www.oup.com/vsi

INNOVATION

A Very Short Introduction

Mark Dodgson & David Gann

This *Very Short Introduction* looks at what innovation is and why it affects us so profoundly. It examines how it occurs, who stimulates it, how it is pursued, and what its outcomes are, both positive and negative. Innovation is hugely challenging and failure is common, yet it is essential to our social and economic progress. Mark Dodgson and David Gann consider the extent to which our understanding of innovation developed over the past century and how it might be used to interpret the global economy we all face in the future.

> 'Innovation has always been fundamental to leadership, be it in the public or private arena. This insightful book teaches lessons from the successes of the past, and spotlights the challenges and the opportunities for innovation as we move from the industrial age to the knowledge economy.'
>
> **Sanford, Senior Vice President, IBM**

www.oup.com/vsi

NOTHING

A Very Short Introduction

Frank Close

What is 'nothing'? What remains when you take all the matter away? Can empty space - a void - exist? This *Very Short Introduction* explores the science and history of the elusive void: from Aristotle's theories to black holes and quantum particles, and why the latest discoveries about the vacuum tell us extraordinary things about the cosmos. Frank Close tells the story of how scientists have explored the elusive void, and the rich discoveries that they have made there. He takes the reader on a lively and accessible history through ancient ideas and cultural superstitions to the frontiers of current research.

> 'An accessible and entertaining read for layperson and scientist alike.'
>
> Physics World

www.oup.com/vsi

NUCLEAR POWER

A Very Short Introduction

Maxwell Irvine

The term 'nuclear power' causes anxiety in many people and there is confusion concerning the nature and extent of the associated risks. Here, Maxwell Irvine presents a concise introduction to the development of nuclear physics leading up to the emergence of the nuclear power industry. He discusses the nature of nuclear energy and deals with various aspects of public concern, considering the risks of nuclear safety, the cost of its development, and waste disposal. Dispelling some of the widespread confusion about nuclear energy, Irvine considers the relevance of nuclear power, the potential of nuclear fusion, and encourages informed debate about its potential.

www.oup.com/vsi

NUMBERS

A Very Short Introduction

Peter M. Higgins

Numbers are integral to our everyday lives and feature in everything we do. In this *Very Short Introduction* Peter M. Higgins, the renowned mathematics writer unravels the world of numbers; demonstrating its richness, and providing a comprehensive view of the idea of the number. Higgins paints a picture of the number world, considering how the modern number system matured over centuries. Explaining the various number types and showing how they behave, he introduces key concepts such as integers, fractions, real numbers, and imaginary numbers. By approaching the topic in a non-technical way and emphasising the basic principles and interactions of numbers with mathematics and science, Higgins also demonstrates the practical interactions and modern applications, such as encryption of confidential data on the internet.

www.oup.com/vsi

TOCQUEVILLE

A Very Short Introduction

Harvey Mansfield

No one has ever described American democracy with more accurate insight or more profoundly than Alexis de Tocqueville. After meeting with Americans on extensive travels in the United States, and intense study of documents and authorities, he authored the landmark *Democracy in America*, publishing its two volumes in 1835 and 1840. Ever since, this book has been the best source for every serious attempt to understand America and democracy itself. Yet Tocqueville himself remains a mystery behind the elegance of his style. In this *Very Short Introduction*, Harvey Mansfield addresses his subject as a thinker, clearly and incisively exploring Tocqueville's writings-not only his masterpiece, but also his secret *Recollections*, intended for posterity alone, and his unfinished work on his native France, *The Old Regime and the Revolution*.

www.oup.com/vsi